Julian Whedon Merrill

Records of the 24th Independent Battery, N.Y.

Julian Whedon Merrill

Records of the 24th Independent Battery, N.Y.

ISBN/EAN: 9783744660143

Printed in Europe, USA, Canada, Australia, Japan

Cover: Foto ©ninafisch / pixelio.de

More available books at **www.hansebooks.com**

He reached Plymouth in time to take part in the battle, and to be taken prisoner.

He died at Andersonville about the same time that his brother died.

155. PRATT, PHILANDER, Perry.—Mustered in at Buffalo, August 31st, 1862.

Was taken prisoner at Plymouth; taken to Andersonville, and died at that place, August 21st, 1864, of chronic diarrhœa.

The number of his grave is 6,455.

Pratt was an excellent cannoneer, ready for duty and quick at his work. A quiet and pleasant comrade. He was one of the useful men at the sawmill at Newport Barracks.

We believe that in the later days at Plymouth, he was promoted corporal.

156. PRINCE, WILLIAM.—Enlisted October 4th, 1864, at Rochester, N. Y., for one year. Joined at Roanoke, November 22d, 1864.

Transferred to Third N. Y. Artillery.

157. PURDY, S. R.—Enlisted at Kingston, N. Y., September 29th, 1864. Transferred to Third New York Artillery, May 25th, 1865. Joined for duty at Roanoke, October 18th, 1864.

158. QUINN, JOHN, Perry.—Joined for duty, November 21st, 1861.

At Washington, where Battery "B," of the Rocket Battalion, was embarking on the vessels for New Berne,

Quinn defended one of the Battery boys who was light and small, in an altercation with a stronger man, a soldier of another regiment. The soldier drew a knife and stabbed Quinn several times; but, notwithstanding this, Quinn continued to fight until he had taken the knife away from his antagonist, and in turn given him several dangerous plunges of the weapon. Upon being separated, both were found to be dangerously wounded, and were removed to the hospital.

Quinn never returned to the Battery.

We have heard that he was residing in Portage.

159. RANKIN, ERASTUS.—Enlisted at Rochester, October 7th, 1864, for one year. Joined at Roanoke, December 1st, 1864. Transferred to Third New York Artillery.

160. RATHBONE, SYDNEY S., Perry.—Enlisted October 3d, 1861. Was discharged some time in 1862, for physical inability.

His historical picture, as represented by the older portion of the Battery boys, was that of a "Jolly old Ambulance driver."

161. RAWSON, PORTER D., Perry.—Enlisted August 26th, 1862.

Mustered in at Buffalo, August 30th, 1862.

Was appointed artificer, November 4th, 1862.

Brought up in a radical school, he believed in showing practically his political tendencies. He left his family and a happy home, to share the privations and the sufferings of his fellows, who were fighting out the principles

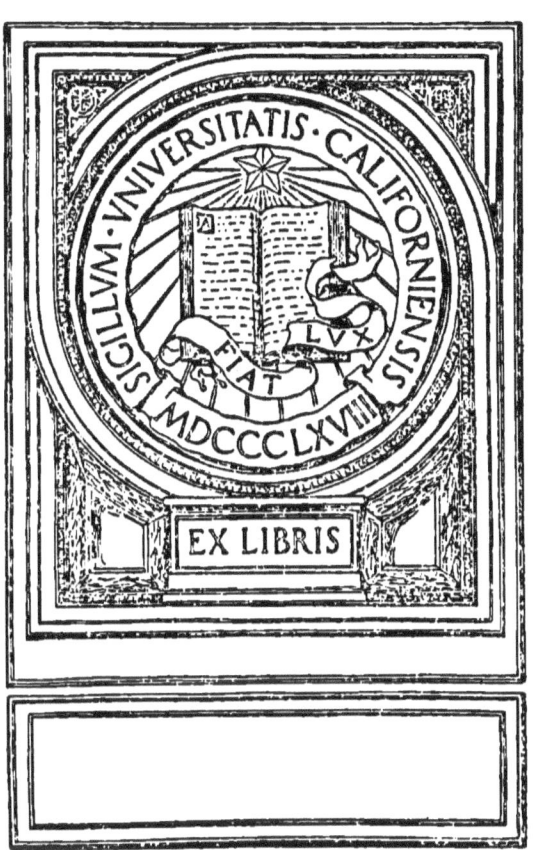

OF THE

24TH INDEPENDENT BATTERY,

N. Y. Light Artillery, U. S. V.

COMPILED BY J. W. MERRILL.

PUBLISHED FOR THE
LADIES' CEMETERY ASSOCIATION, OF PERRY, N. Y.

1870.

J. O. SEYMOUR, KENNARD & HAY,
Printers and Stationers,
89 Liberty Street, New York.

CONTENTS.

PART I.
INTRODUCTION.

COMPANY ROLL.

List of names of members who were killed in battle—List of names of members who died at Andersonville—List of names of members who died at Florence—List of names of members who died at Charleston—List of names of members who died after reaching Federal Lines—List of names of members who died of disease in United States Hospitals—List of names of members who died at their homes (while yet in the United States service)—Official roster of the Rocket Battalion—Official roster of the Twenty-fourth New York Battery—List of promotions of members of the Twenty-fourth New York Battery—List of volunteers from Perry attached to other army organizations.

PERSONAL SKETCHES.

1. Members of Twenty-fourth New York Battery. 2. Other volunteer soldiers from Perry.

PART II.

CHAPTER I.
THE ROCKET BATTALION.

CHAPTER II.
THE ROCKET.

CHAPTER III.
BATTERY B.

CHAPTER IV.
THE TWENTY-FOURTH NEW YORK INDEPENDENT BATTERY.

CHAPTER V.
NEWPORT BARRACKS.

CHAPTER VI.
KINSTON, GOLDSBORO' AND WHITEHALL.

CHAPTER VII.
NEW BERNE.

CHAPTER VIII.
PLYMOUTH.

CHAPTER IX.
THE BATTLE OF PLYMOUTH.

CHAPTER X.
THE CAPTURE OF PLYMOUTH.

CHAPTER XI.
ANDERSONVILLE.

CHAPTER XII.
FACTS AND THEORIES.

APPENDIX.

INTRODUCTION.

The object of publishing this work is:—

1st—To place in permanent record the experiences, adventures and sufferings of that brave band of young men, who composed the organization entitled the Twenty-fourth New York Independent Battery of Light Artillery.

2d—To prompt the reader to the fact, that the services of all these patriotic men, who left their families and their firesides, to battle for their country, who pledged and offered up their lives, that you might continue life in peace, prosperity and freedom, have never been in a public manner, properly recognized.

Although the memory of those who died while engaged in suppressing the Great Rebellion is more particularly dear to their relatives and friends, yet, at the same time, it is cherished by all loyal people.

No County in the United States was more loyal than Wyoming County—none more enthusiastic, none more generous, when they started the boys off to the wars.

No County has better reasons for being proud of its representatives in the war.

None has had greater cause to mourn.

Is there less of that sentiment of loyalty in their hearts to-day? Is there less of generosity? Is there no gratitude? Can they forget so soon?

Let a practical answer to these questions be a strong endeavor to purchase and erect a monument, in some sightly position, at the sight of which, passers-by in our times shall say—"These people do their noble dead a deserved honor." And future generations shall learn from it of the heroic deeds and of the sufferings of the young men of this generation, who helped to fight the battles that finally won for us and for them a country freed from oppression, tyranny and wrong.

INTRODUCTION.

In our families we grieve over the loss of any dear member; it is a gratification to us to show our love and our respect by rearing marble slabs and shafts, keeping in decency and order their burial grounds, and decorating their graves with plants and flowers. Far away in sunny Southern lands, and just so far away from loving hearts and hands, lies one who is as dear to you, perhaps, as any who sleep in your own home burial ground.

A number on a whitened board is all that marks each individual resting place. Shall that remain the only remembrance of their sufferings, their heroism, and their death?

I say again, injustice is done the dead, who did these noble deeds, and who suffered these appalling tortures.

Their simple story is eloquence. Let its pathos be a power to influence you to respond with heart and purse, to the call to acknowledge their heroism and their offices, by erecting to their memory a monument, which, like their heroic acts, shall be perfect, be grand, be forever.

Act as enthusiastically, and give as generously, as you did when they responded to their country's call.

I would particularly call the attention of the surviving members of the Battery to the object for which this book has been written and published, and ask them to give aid towards the accomplishment of this sacred undertaking.

All of you have passed many pleasant hours with these, our comrades, whose deeds we would memorize. In earlier days, they were our playmates and schoolmates. In later days, we were in friendly contest for the honors of college, or in rivalry with them in mercantile pursuits.

Finally, under the loved banner of our country, hand in hand we battled for the right and victory.

By the kindly interpositions of Divine Providence we live; by the decree of the same power they were called away from us and from all that was dear to them on earth.

Let us then, as we esteemed and honored them in life, continue so to do, in such honorable death.

The compilation of this work, neither demands nor affords opportunity or place for a display of composition, words, theories, ideas, or of any of the properties which are considered essential to perfect and beautify a book. It resolves itself into a simple narrative of the pleasant and sad experiences of one small band of men among many thousands such, that participated in putting down the Great Rebellion. It is collected from the records at the War Department, correspondence of local papers, private diaries and private correspondence. It consequently, naturally assumes a tone of familiarity.

It is of interest to but few; but to that few, how intensely interesting. To them, I believe, the story will be most attractive and most satisfying in its simplest garb; the facts that appear in these pages, need no touch of fancy's brush, to bring out a vivid scene of despair and sorrow.

Have we not all in our younger days, when the mild Spring atmosphere has converted the crusted snow into a surface resembling thin sheets of cotton, participated in the merry sport of rolling up huge snow balls? The handful of snow, rounded by hardy little fists, dropped into the adhering mass, was pushed along, and gradually assumed larger and larger proportions, until it increased to a magnitude satisfactory to the wishes of the juvenile laborers. Then it was wondered at and admired, and each little mind was proud of the portion of the work it had done towards bringing it to such perfection. But another day came, the sun poured down its rays, and the structure quickly, and almost invisibly, melted away. In tracing its history, we find such to be comparatively the beginning, the perfection and the annihilation of the organization called the Twenty-fourth New York Battery. In October, 1861, a handful of Perry boys enlisted in the service of the United States as Artillerymen. To this nucleus was gradually added recruits, until finally, in December, 1863, it was a complete Six Gun Battery, of 125 men. How it worked, and what it did during its eventful career, from its entrance to its exit, is told in the following pages.

A warm Spring day was April 20th, 1864; the enemy came in overpowering numbers, the Company was captured, and silently, gradually, the members of the organization passed away, until a little remnant was mustered out of service, at Albany, in the Spring of 1865. As men, we had but reproduced our boyhood's play, and worked hard to roll the ball to a satisfactory shape and size, only to see it quickly and like a vapor pass away.

COMPANY ROLL,
24th New York Battery of Light Artillery.

OFFICERS:

LEE, J. E.
CADY, LESTER A.

HASTINGS, GEORGE S.
GRAHAM, GEORGE W.

PRIVATES.

1 Adams, Abner
2 Ainsworth, R. C.
3 Ainsworth, William
4 Alburty, Wm.
5 Alburty, F. M.
6 Allen, Z.
7 Andrus, Lemuel
8 Andrews, Mark
9 Armstrong, Wm.
10 Armstrong, J. H.
11 Atwood, George S.
12 Ausbacher, Moses

13 Baker, John
14 Barker, Gustavus
15 Barnes, Roswell H.
16 Bartley, John
17 Bartlett, Hartwell
18 Bachelder, B. F.
19 Beers, L. M.
20 Birdsall, George
21 Billingham, Ira
22 Blood, Wm.
23 Blake, W. D.
24 Boies, Edwin
25 Brooks, John A.
26 Brayton, Rufus
27 Brown, George
28 Bullock, Robert
29 Bulkley, W. E.
30 Bulkley, Chas.
31 Buck, Robert
32 Button, James
33 Burd, H. C.

34 Calhoun, G. W.
35 Calteaux, Paul
36 Calkins, James
37 Camp, William S.
38 Camp, George
39 Canfield, S. D.
40 Carnahan, William
41 Carnahan, Charles
42 Chapman, John
43 Chadbourne, Henry
44 Chapin, William E.
45 Clark, C. A.
46 Clute, H. V.
47 Comstock, A. W.
48 Cook, Harlo
49 Corbin, B. F.
50 Corkwell, John
51 Cowen, James
52 Crooker, W. W.
53 Crooks, J.
54 Crosby, Morton
55 Crounce, George
56 Cusick, Hiram
57 Culver, A. L.
58 Cypher, George W.

59 Davis, Ornau
60 Dolbeer, C. H.
61 Duryea, George
62 Duryea, Joseph

63 Eastwood, Edwin

64 Farrell, Philemon

COMPANY ROLL.

65	Ferrin, J. T.	108	Lapham, L. H.
66	Ferguson, Andrew T.	109	Lapham, Horace
67	Filbin, John	110	Lawler, E.
68	Finnigan, Dennis	111	Lee, Abram
69	Fitch, Charles W.	112	Lent, Abram
70	Fitzgerald, Thomas	113	Leonard, Francis
71	Fitzpatrick, Pierce	114	Lloyd, H. P.
72	Flynn, James	115	Loomis, Hiram
73	Foster, Henry		
		116	McClair, Jerry
74	Galusha, J. E.	117	McCrary, Orrin S.
75	Goodhue, D. W.	118	McCrary, Wm. A.
76	Gould, Willard	119	McCrary, Charles
77	Grant, Murray	120	McCrink, John
78	Green, Lawrence	121	McCrink, James
79	Griffith, Charles R.	122	McDonald, Arch'd.
80	Griffith, Albert	123	McEwen, Geo. W.
81	Grisewood, Thomas	124	McGuire, Thos.
		125	McGuire, James
82	Hart, Charles	126	McGuire, Michael
83	Harmon, John C.	127	McNinch, Henry
84	Harrington, M.	128	McVey, James
85	Hastings, Fred. E.	129	Marean, Chas. A.
86	Hathaway, Charles	130	Marrin, Patrick
87	Hinton, W. H.	131	Marrin, Connor
88	Holman, George A.	132	Martin, H. C.
89	Hollister, Benjamin H.	133	Meade, Geo. F. H.
90	Homan, Chas. H.	134	Merrill, J. W.
91	Horton, Chas.	135	Miller, George
92	Hosford, W. F.	136	Miner, J. Gile
93	Hoyt, Wilbur M.	137	Mosier, Marion R.
94	Hubbard, H.	138	Munroe, Darius
95	Hughson, Wallace E.	139	Murray, W. R.
96	Humphrey, Arthur		
97	Humphrey, Chas.	140	Newcomb, L.
98	Hunter, E. H.	141	Newton, Riley J.
99	Hurlburt, E. T. M.	142	Nichols, Samuel
		143	Nichols, William P.
100	Jackson, Dan'l.		
101	Johnson, Geo. B.	144	Otis, F. D.
		145	Otis, Chas.
102	Keeney, George W.	146	O'Dell, Thos.
103	Keith, G. H.		
104	Kellogg, Geo. W.	147	Page, H. C.
105	Ketchum, R. A.	148	Page, Wm. N.
106	King, Sylvanus	149	Parmlee, O. G.
107	Knowlden, Henry C.	150	Patterson, Wm.

COMPANY ROLL.

151 Perkins, Jas. W.
152 Phelan, Chas. T.
153 Piper, Geo. W.
154 Piper, A.
155 Pratt, Philander
156 Prince, William
157 Purdy, S. R.

158 Quinn, John

159 Rankin, Erastus
160 Rathbone, Sydney S.
161 Rawson, Porter D.
162 Raymond, Henry
163 Rich, Thurmon
164 Richards, Elias
165 Richards, Albert
166 Richardson, Orlando
167 Roach, Wm.
168 Rood, Le Grand D.
169 Root, Hiram
170 Root, Stephen
171 Rowell, Solon
172 Russell, Enoch J.
173 Russell, John A.
174 Russell, John

175 Sackett, Walter
176 Stafford, Pembroke J.
177 Sanford, L. J.
178 Secor, A. J.
179 Shank, Laban H.
180 Shell, John
181 Sheppard, Nelson
182 Shirley, Phares

183 Shockensey, Timothy F.
184 Smith, Mason C.
185 Smith, J. W.
186 Stevens, Geo. W.
187 Stoddard, Samuel
188 Storms, Thomas S.
189 Sunderland, Charles
190 Sunfield, James

191 Thayer, Lewis P.
192 Tilton, Henry
193 Tirrell, Samuel
194 Truair, O. M.
195 Turner, Robt.

196 Van Buren, Sylvester

197 Wardwell, E. H.
198 Washington, George
199 Wayne, Joseph
200 Welch, Edward
201 Weller, Jacob H.
202 Wetmore, Chauncey
203 Whitney, Hamilton S.
204 Whitney, W. A.
205 Whitbeck, Henry
206 Williams, Oliver
207 Williams, Thos.
208 Winne, B. V. L.
209 Wood, Emmett
210 Woolsey, John
211 Woolsey, Etting
212 Wright, George G.

213 Yancer, J. D.

LIST OF DEATHS.

List of Names of Members who were killed in Battle.

Fitzpatrick, Pierce
Hoyt, Wilbur M.
Meade, Geo. F. H.
Turner, Robert

List of Names of Members who died at Andersonville Prison.

Alburty, Wm.
Armstrong, Wm.
Atwood, George S.
Baker, John
Barnes, Roswell
Bartlett, Hartwell
Bachelder, B. F.
Blake, W. D.
Button, James
Calteaux, Paul
Calkins, James
Carnahan, Chas.
Chadbourne, Henry
Clute, H. V.
Comstock, A. W.
Corbin, B. F.
Crosby, Morton
Crounce, George
Culver, A. L.
Eastwood, Edwin
Filbin, John
Fitch, Chas. W.
Fitzgerald, Thos.
Flynn, James
Griffith, Chas. R.
Griffith, Albert

Hathaway, Chas.
Hosford, W. F.
Hunter, E. H.
Johnson, Geo. B.
Keeney, Geo. W.
King, Sylvanus
Lapham, L. H.
Lee, Abram
Lent, Abram
McCrink, John
McDonald, Arch'd
Marean, Chas. A.
Martin, H. C.
Miner, J. Gile
Newton, Riley J.
Pratt, Philander
Rich, Thurmon
Rood, Le Grand D.
Safford, Pembroke J.
Shank, Laban H.
Shirley, Phares
Shockensey, Timothy F.
Smith, Mason C.
Tilton, Henry
Welch, Edward
Williams, Oliver

Wood, Emmett

List of Names of Members who died at Florence Prison.

Bartley, John
Blood, William
Brooks, John
McCrary, Orrin S.
McCrink, James
McNinch, Henry
Piper, Geo. W.
Piper, A.
Root, Stephen
Stevens, Geo. W.
Tirrell, Samuel
Wetmore, Chauncey

List of Names of Members who died at Charleston Prison.

Ainsworth, William
Rawson, Porter D.

List of Names of Members who died after reaching the Federal Lines.

Galusha, J. E.
Nichols, William P.
Nichols, Samuel

List of Names of Members who died in United States Service, of Disease.

Andrus, Lemuel
Beers, L. M.
Brayton, Rufus
Grant, Murray
Keith, G. H.
McCrary, Wm. A.
McGuire, Michael
Munroe, Darius
Otis, F. D.
Truair, O. M.

List of Names of Members who died at their Homes (while yet in the United States Service).

Billingham, Ira
McVey, James

(*From Official report of the Adjutant-General of the State of New York*, 1868.)

Rocket Battalion.*

This battalion was raised and organized at Albany, N. Y., to serve three years. It was mustered into the service of the United States, December 6, 1861, and changed to the Twenty-third and Twenty-fourth Independent Batteries New York Artillery, February 11, 1863.

NAME.	Date of Commission.	Date of Rank.	Remarks.
Major:			
Thomas M. Lyon	Dec. 9, 1861	Dec. 7, 1861	Discharged June 28, 1862.
Captains:			
Alfred Ransom	Dec. 9, 1861	Nov. 12, 1861	Transferred to 23d Independent Battery Feb. 11, 1863.
Jay E. Lee	Dec. 9, 1861	Oct. 26, 1861	Transferred to 24th Independent Battery Feb. 11, 1863.
First Lieutenants:			
Henry W. Dodge	Dec. 9, 1861	Dec. 7, 1861	Resigned April 27, 1862.
Samuel Kittinger, Jr..	May 13, 1862	Apr. 27, 1862	Transferred to 23d Independent Battery Feb. 11, 1863.
George S. Hastings	Sept. 13, 1862	Aug. 30, 1862	Transferred to 24th Independent Battery Feb. 11, 1863.
Lester A. Cady	Dec. 9, 1861	Oct. 26, 1861	Transferred to 24th Independent Battery Feb. 11, 1863.
Second Lieutenants:			
Samuel Kittinger, Jr..	Dec. 9, 1861	Nov. 12, 1861	Promoted to First Lieutenant May 13, 1862.
Thomas Law	May 13, 1862	Apr. 27, 1862	Transferred to 23d Independent Battery Feb. 11, 1863.
Charles C. T. Keith	Mar. 10, 1862	Jan. 2, 1862	Transferred to 23d Independent Battery Feb. 11, 1863.
George M. Graham	Dec. 9, 1861	Dec. 7, 1861	Transferred to 24th Independent Battery Feb. 11, 1863.
Frederick E. Hastings	June 17, 1862	June 2, 1862	Transferred to 24th Independent Battery Feb. 11, 1863.

* Changed to the Twenty-third and Twenty-fourth Independent Batteries of Artillery, New York Volunteers, by S. O. No. 81, A. G. O., Albany, February, 11, 1863.

(*From Official report of the Adjutant-General of the State of New York*, 1868.)

Twenty-fourth Battery Light Artillery.

This Battery (formerly Company B, Rocket Battalion) was organized at Albany, N. Y., to serve three years. It was raised, principally, in the counties of Monroe and Wyoming. Mustered into the service of the United States, December 6, 1861. On the expiration of its term of service the original members (except veterans) were mustered out, and the veterans and recruits transferred to the Third Regiment New York Artillery, March 8, 1865.

KINSTON; WHITEHALL; GOLDSBORO'; NEW BERNE; PLYMOUTH.

NAME.	Date of Commission.	Date of Rank.	Remarks.
Captains:			
Jay E. Lee............... (*Brevet Lt.-Col. N. Y. V. & Brevet Lt.-Col. U.S.V.*)	Dec. 9, 1861	Oct. 26, 1861	Resigned June 13, 1863.
Lester A. Cady..........	June 23, 1863	June 13, 1863	Discharged Dec. 29, '64.
William W. Crooker......	Jan. 28, 1865	Jan. 10, 1865	Not mustered.
First Lieutenants:			
George S. Hastings...... (*Brevet Colonel N. Y. V.*)	Sept. 13, 1862	Aug. 30, 1862	Discharged Jan. 4, 1865.
William S. Camp.........	Jan. 28, 1865	Dec. 28, 1864	Discharged March 8, '65.
Lester A. Cady..........	Dec. 9, 1861	Oct. 26, 1861	Promoted Captain June 23, 1863.
Frederick E. Hastings... (*Brevet Captain N. Y. V. & Brevet Major U. S. V.*)	June 23, 1863	June 13, 1863	Discharged Jan. 22, '65.
Second Lieutenants:			
George W. Graham.......	Dec. 9, 1861	Dec. 7, 1861	Transferred to 3d N. Y. Cavalry, Dec. 24, 1863.
Edward H. Wardwell.....	Apr. 24, 1863	Apr. 15, 1863	Resigned Aug. 30, 1864.
Lucius S. Newcomb......	Jan. 28, 1865	Jan. 10, 1865	Not mustered.
Frederick E. Hastings....	June 17, 1862	June 2, 1862	Promoted to First Lieutenant June 23, 1863.
Charles H. Dolbeer........	June 23, 1863	June 13, 1863	Discharged Jan. 22, '65.
*Chas. F. W. F. De Werner	......	......	Resigned May 31, 1862.

* On records of War Department; not commissioned.

List of Promotions to Commissions.

NAME.	Date of Commission.	Remarks.
Major:		
Lloyd, H. P...................	Jan. 24, 1865	Twenty-second New York Cavalry.
Captains:		
Cady, Lester A...............	June 23, 1863	Twenty-fourth New York Battery.
Canfield, S. D................		
Crooker, W. W...............	Jan. 28, 1865	Twenty-fourth New York Battery.
Graham, George W...........	Dec. 9, 1861	Third New York Cavalry.
Lee, J. E......................	Dec. 9, 1861	Twenty-fourth New York Battery.
(*Brevet Lt.-Col. N. Y. V. & Brevet Lt.-Col. U. S. V.*)		
Lloyd, H. P...................	July 13, 1864	Twenty-second New York Cavalry.
First Lieutenants:		
Andrews, Mark...............	Sept. 10, 1863	Twentieth New York Battery.
Camp, William S.............	Jan. 28, 1865	Twenty-fourth New York Battery.
Canfield, S. D................		
Clark, C. A....................	July 3, 1865	Twentieth New York Battery.
Cady, Lester A...............	Dec. 9, 1861	Twenty-fourth New York Battery.
Hastings, George S...........	Sept. 13, 1862	Twenty-fourth New York Battery.
(*Brevet Colonel N. Y. V.*)		
Hastings, Fred. E.............	June 23, 1863	Twenty-fourth New York Battery.
(*Brevet Captain N. Y. V. & Brevet Major U. S. V.*)		
Kellogg, George W...........	Oct. 31, 1864	Third New York Battery.
Lloyd, H. P...................	July 12, 1864	Twenty-second New York Cavalry.
McVey, James................	Aug. 31, 1863	Third New York Artillery.
Newcomb, Lucius S..........	July 5, 1865	Third New York Artillery.
Wardwell, E. H...............		
Second Lieutenants:		
Adams, Abner................	Feb. 20, 1863	First N. C. U. V.
Clark, C. A....................	Mar. 29, 1864	Sixteenth New York Artillery.
	Feb. 1, 1865	Twentieth New York Battery.
Dolbeer, C. H.................	June 23, 1863	Twenty-fourth New York Battery.
Graham, George W...........	Dec. 9, 1861	Twenty-fourth New York Battery.
Hastings, Fred. E.............	June 17, 1862	Twenty-fourth New York Battery.
Kellogg, George W...........	Jan. 8, 1864	Third New York Battery.
McClair, Jerry................		
McVey, James................	Aug. 31, 1863	Third New York Artillery.
Merrill, J. W..................	April 20, 1864	Second New York Artillery.
Newcomb, Lucius S..........	Jan. 28, 1865	Twenty-fourth New York Battery.
Page, Wm. N..................		Fourth Artillery.
Wardwell, E. H...............	April 24, 1863	Twenty-fourth New York Battery.

PERSONAL SKETCHES.

———•◆•———

After the compilation of this work was deemed advisable by the writer and his friends, the following Circular was sent to every member of the Battery whose address could be ascertained :

NEW YORK, January 21st, 1869.

MY DEAR SIR:

For some time I have been employing my leisure hours in searching for matters and documents relating to the " TWENTY-FOURTH NEW YORK BATTERY."

I want to possess a complete record of its entire progress from its organization up to the time it was disbanded, and the further personal history of each individual member to the present time.

I find I have undertaken a difficult task, and I am obliged to ask your assistance. Will you not oblige me, therefore, by answering the following questions:—

When and where did you enlist ?
When and where were you mustered in ?
When and where were you mustered out ?
Did you re-enlist as a Veteran ?

Were you in any Skirmishes or Battles? If yes, when and where?

Were you in any Southern Prisons? If yes, what Prisons? Give particulars of imprisonment.

Do you positively know of any of your Comrades dying? If yes, give name, date and place.

When and where were you paroled and exchanged?

What did you do after you were paroled until you were mustered out?

Were you promoted while in the army? If yes, when and to what positions?

Give Non-commissioned and Commissioned Appointments, with full particulars.

Were you on any detailed duty? If yes, where, when, and in what position?

Give residence and occupation, since your retirement from the army.

Relate any incident in the army about yourself, or your friends, which you think of interest.

Have you married since your enlistment? If yes, when, and to whom?

What is your present address?

Should fortune favor, I shall publish a record of the doings of our Battery. I hope you will render me this assistance.

Yours, &c.,

Wherever in this book a Personal Sketch is not full, it occurred from our being unable to obtain information.

It will be noticed that in the "Personal Sketches," there is no mention of a large number of reductions of Non-Commissioned Officers.

The Battery passed through several changes of organization, and of administration. As a rule, differences of opinions and personal considerations, influenced and brought about these reductions.

At the time of the Mutiny, and at other times, Non-Commissioned Officers asked to be reduced. Only in two or three cases of reduction, are we aware that there was any reflection upon the soldierly qualities of the officer reduced.

Space could not be spared to give a detailed history of each case—allowing both sides of the story to be told—in this book.

The writer trusts and believes that the surviving members of the Battery will understand these remarks; and that they will consider these reasons ample for not reviving questions of differences, by particularizing those reductions.

There were indeed few left to whom the Circular letter could be addressed, but those who did respond gave us much valuable information. In many cases, friends and relatives replied. In all cases where it was possible, we have given the sketch of each member, as it was furnished us, either by himself or his friends. In some instances, however, the writer knew more of the experiences of his comrades than any other person; and in such cases he has himself taken the liberty to sketch the character and experiences of his comrades.

He would here take the opportunity of thanking the following persons for their aid, and for the information which they so pleasantly and cheerfully furnished.

Mrs. C. A. Cleveland,	Lucius S. Newcomb,
Mrs. A. D. Keeney,	Wm. R. Murray,
Miss Mary Smith,	H. C. Burd,
Jay E. Lee,	George Birdsall,
George S. Hastings,	Andrew Ferguson,
C. A. Clark,	H. P. Lloyd,
C. H. Dolbeer,	Charles Homan,
Wm. S. Camp,	R. D. Higgins.

There were many others who encouraged him by writing agreeable responses; and they too will please accept his heartiest thanks.

The Personal Sketches are placed in the same order as the Company Roll.

PERSONAL SKETCHES.

LEE, JAY E., Captain.—In the fall of 1861, Mr. Lee, a young and successful lawyer of Perry, convinced that more men were needed in our army, determined to offer his services to the Government. Upon investigating the tactics of the different branches of the service, he selected the artillery, as that most needed and most desirable. Together with Mr. Wyckoff, he set about interesting the young men of the place, in organizing a company which should represent that town and vicinity in the great conflict.

About fifty men signed a declaration to join the company, but from some cause, when they were called upon to proceed to Buffalo and muster in, only about twenty responded. These consolidated with other squads from different towns, and formed a company, of which Mr. Lee was elected captain. He was mustered into the service October 26th, 1861, and commissioned as captain in the "Rocket Battalion," December 9th, 1861. Soon after he was commissioned, the people of Perry, appreciating his efforts and ability, presented him with a purse, an account of which is given in the "Wyoming Times" of February 7th, 1862, as follows:

"SOMETHING MORE THAN A COMPLIMENT.—The following correspondence explains itself, but we cannot refrain from accompanying it with the expression of our gratification, in view of the handsome and spirited manner in which the thing was done—quite characteristic of our community, by the by—and the peculiar propriety of doing it. Captain Lee moved in the matter of organizing a company when our national affairs were in their most gloomy condition; greatly influenced, as we happen to know, by the consideration that his country needed his services. * * * *

"'PERRY, January 29, 1862.
"'Capt. J. E. LEE:
"'*Sir:*—In behalf of a large number of your fellow-citizens, residents of Perry and vicinity (a list of whom I enclose), I have the honor of presenting to you the enclosed sum of sixty-five dollars, contributed by them for the purpose of purchasing side arms (sword and pistols), for your use in the service of your country, to which you have so generously and at so great a sacrifice devoted yourself.

"'Please accept it, to quote the language of the subscription, "as an expression of their appreciation of your patriotic and successful efforts in raising your company, and of your admirable fitness to command it." Trusting that the efforts of the Government and the people to put down this unrighteous rebellion may speedily be crowned with success, and that you soon may be restored to your family and friends.
"'I am, yours respectfully,
"'H. N. PAGE.'

"'PERRY, February 1, 1862.
"'H. N. PAGE, Esq.:
"'*Dear Sir:*—Your communication, enclosing sixty-five dollars for the purchase of side arms for my use, with a list of the donors, is just received. Permit me, through you, to thank my friends for this handsome gift, and the flattering words with which they grace their generosity. * * * *

"'With you, I earnestly hope this unrighteous rebellion may speedily be crushed, and that I, with others, who are self-banished from our pleasant homes, may soon be permitted to return to your midst; and, moreover, when I do return, that no act of mine while wearing these arms, shall cause me to avoid the greeting of any of whose generosity and patriotism they are the indices.

"'With genuine gratitude, I am,
"'Very truly and respectfully,
"'Your obedient servant,
"'J. E. LEE.'"

Capt. Lee was in command of the section that participated in the battles of Kinston, Goldsboro' and Whitehall. He was also in command of the Battery at the second attack on New Berne, N. C.

While in the service, he was attacked with a severe hemorrhage of the lungs, the result of exposure and over exertion, which unfortunate event compelled him to resign his commission, while stationed at Plymouth, N. C., June 13, 1863. The following letter was read to the Battery, after his resignation had been accepted:

"NEW BERNE, N. C., June 13, 1863.
"TO THE MEMBERS OF THE TWENTY-FOURTH N. Y. BATTERY:

"I can no longer address you as "my men," or "fellow soldiers," but I can say what is as good or better, my friends, you have just heard the order which discharges me from the service of the United States, and sunders my connection with the dear old Twenty-fourth. I regret exceedingly that I cannot see you all again, and say good bye with my own voice, and give each a parting grasp of the hand. When I left you last Thursday, I had already prepared my resignation, and did not expect to return, and it made me feel very badly to come away without saying good bye; but I could not; my resignation had not been acted upon, and we have all learned that nothing is certain in military matters but uncertainty. I had no right to take it for granted that it would be accepted, so I was obliged to leave in silence, as I did. My motives and actions cannot be misconstrued, however, I trust no one of you will be so ungenerous as to think I desired to steal away from you.

"One of my greatest regrets at leaving the service is, that I cannot in person take my leave of you, and assure each one of my high regard and lasting attachment. As it is, let me say that there is not a man in the Battery who has not a firm hold on my memory and heart. I shall constantly carry with me the deepest interest and anxiety for the Battery, and every individual member of it; and not only while you are in the service, but as long as you or I shall live. If length of years is given me or you, I know that in after life when we meet, warm and earnest will be our words of greeting, and a thrill of pleasure will follow the hearty grasp of our hands. I have not only deep personal attachment for all of you, but I am proud of you as an organization. If I was to remain in the service I should want no other command. Nothing would tempt me from you. My ambition has not been for promotion, but to make you thorough or efficient soldiers. In that I trust I have succeeded. I sincerely believe that in all the armies of the United States, there cannot be found an organization better fitted to do thorough and earnest work for its country, than the Twenty-fourth New York Battery. I am not so vain, however, as to take to myself all the credit for this; I have had able and willing assistance in my officers, from highest to lowest, and above all, I have had intelligent, honorable, *manly* men to govern and instruct. No other organization of like size in the army can boast a tithe of the intelligence, education, and high-toned manliness and moral character this company contains. Under such circumstances I had shown myself but poorly fitted for the position I have had, if now, I had not a battery to be proud of. I have been proud of my command, and shall let no opportunity go unimproved of boasting of it when I am home among you and my friends.

"This, at least ought to be granted me, as I have no deeds of valor, &c., to boast of. Of my reasons for resigning my position and leaving you, there can scarcely be any necessity of speaking. I presume it was not unlooked for by any one, certainly not by those who were familiar with my physical condition. For more than a year I have been unfit for military service, and I considered it my duty to resign a position whose duties I could not perform. The officers who remain are tried and true. You know them well, and, I believe, have full confidence in them. Upon you, I know they rely.

"Let me now say a final good bye, and God bless and preserve you.
  "Your friend and former Captain,
                    "J. E. LEE."

The Battery went through many changes, and Capt. Lee was not always popular. In Washington, D. C., March, 1862, he was tried before court martial, on several charges, but was honorably acquitted. Still, we think that even those who did not accord with his views and decisions, would admit that he was an able and efficient officer; and at the time he resigned, the Battery was one of the finest appearing and best drilled batteries in that department. He was brevetted major and lieutenant colonel of United States Volunteers, for " gallant and meritorious conduct," and also received a commission as brevet lieutenant colonel of New York Volunteers. Soon after his discharge from service he visited the Western Territories, California and the Sandwich Islands, for the purpose of benefitting his health. In January, 1866, he received an appointment on the staff of Governor Fenton, with rank of lieutenant colonel, and was assigned to duty as military agent for the State of New York, at Washington.

He is at present at Jacksonville, Fla., where he is practicing law.

CADY, L. A., Captain.—Enlisted in Hamlin, Monroe Co., N. Y., in October, 1861, and upon the organization of the Battery, was mustered in as its second lieutenant. He remained in active service until December, 1864, rising to the rank of captain. He participated in every battle in which the Battery was engaged, evincing the qualities of the good soldier. In entering the army he was actuated by a noble patriotism that led him to make great personal and domestic sacrifices, with cheerful alacrity, thoroughly comprehending the mighty issues of the long struggle ; he always had a stanch faith in the integrity of the cause, and an unwavering confidence in its ultimate triumph.

He was a faithful and diligent officer, with a quick appreciation of the fidelity of the humblest member of the Battery; and a just pride in the intelligence, good discipline and splendid appearance of his command. He was captured in the battle of Plymouth, and suffered the rigors of prison life, until early in October, 1864, when he escaped, while *en route* from Charleston to Columbia, S. C.

After a weary experience in the swamps, forests and mountains of the South, he reached the Union lines at Strawberry Plains, in East Tennessee. His health was considerably impaired by the hardships and exposure he had undergone; and there being but a fragment of the Battery for duty, he was induced to resign his commission.

His health continued to fail, and though his strong love of an active and useful life, and the devotion of an affectionate wife and children, furnished the most powerful motives for living, his hitherto tireless energies were destined to succumb to the fatal force of disease. He died on the 8th day of November, 1865, at Waterford, Orleans Co., N. Y.

HASTINGS, GEORGE S., First Lieut.—In the autumn of 1861, he removed from Oswego to Perry, N. Y., where he engaged in the practice of law, until August, 1862. In common with thousands of patriotic young men, he then believed that duty called him to his country's service. He accordingly procured authority to recruit for the Twenty-fourth New York Battery, then known as Battery " B," of the " Rocket Battalion." Among the pioneer recruits who gave him cheerful and patriotic co-operation, were Mason C. Smith, Phares Shirley, Oliver Williams, William S. Camp, Charles Dolbeer, and J. W. Merrill.

The work of enlistment was sharp, short and decisive; commencing in earnest on the 25th of August, and receiving a strong impetus at a public meeting held in Perry, on the evening of the following day. In the same week sixty-four young men had enlisted; sixty of whom were accepted and mustered in at Buffalo, on August 30, 1864. Shortly afterwards, the detachment, numbering about seventy, joined the battery, in North Carolina. Returning to Perry to make necessary and final arrangements to follow his comrades, after a brief interval, he joined the command at Newport barracks. In March, 1863, the Battery was ordered to Plymouth, then an insignificant station on the Roanoke. It afterwards became one of an important line of fortified posts on the coast region of North Carolina, and was the head-quarters of the sub-district of the Albemarle. While there, Lieutenant Hastings was detailed as Judge-Advocate of the sub-district, and served in that capacity until the battle of Plymouth. Having been taken prisoner, he was sent to Macon, Ga., remaining there until August, 1864. During this time he made two attempts to escape, both of which were defeated at the point where success seemed assured. In August, while on the route from

Macon to Charleston, he escaped from the cars; and after a wearisome, painful and solitary tramp of four nights, was pursued by dogs and recaptured. His citizen captors were disposed to regard him as a spy, and for a time he had an unpleasant foretaste of the pains and penalties visited upon curious intruders.

The Commanding General of the Department of Georgia was pleased to restore him to the rigors, facetiously styled, "The rights of a prisoner of war." A month in sultry Savannah; a fortnight in the filthy jail yard of Charleston, with pestilential odors below, and screaming shells above; and five days in the prison camp near Columbia, S. C., rounded the period of his probation as the unwilling partaker of "Southern Hospitality." Then a night escape through the cordon of sentinels who guarded the camp; a long pilgrimage through the pines of the Palmetto State, and over the rugged mountains of North Carolina and Tennessee; and the dream of liberty was realized under the flag whose folds were so dear to the sturdy loyalists of East Tennessee. Shortly afterwards he received an appointment upon the staff of Governor Fenton, and subsequently became his private secretary, in which capacity he remained in Albany, until September, 1868. He then removed to New York, and held the position of Assistant Attorney of the Board of Excise.

He is now practicing his profession at 4 and 6 Pine Street, New York.

GRAHAM, GEORGE W.—Was commissioned as a second lieutenant in the "Rocket Battalion," December 9th,

1861. Transferred to the Twenty-fourth New York Battery, February 11th, 1863. Transferred to the Third New York Cavalry, December 24th, 1863. From thence transferred to the First North Carolina Cavalry. He was a dashing and reckless officer. Several of his reported exploits exhibited both coolness and audacity. At the reorganization of the army he was appointed first lieutenant in the United States Army; has since been promoted to captain of the Tenth Cavalry, and is now with his regiment, somewhere on the borders.

1. ADAMS, ABNER.—Enlisted at Albany, September 30th, 1862, and mustered in there, October 1st, 1862. He was discharged at New Berne, February 20th, 1863, and promoted to second lieutenant in the First North Carolina Union Volunteers. Held the position of Military Secretary to the Military Governor of Department of North Carolina, (Governor Stanly), and resigned June 10th, 1863, to leave the army. In September, 1864, he re-entered the army, as private, in the Twenty-fourth New York Battery; was on detached service, as clerk in Provost Marshal's office of Twenty-fourth District, New York, and mustered out in June, 1865.

Married Miss M. E. French, in Livingston Co., N. Y., 21st October, 1863, and has one child, named "Robert Turner Adams," to perpetuate the name of a member of the Battery, who was killed in North Carolina, in the fall of 1862.

His present address is Rochester, N. Y.

2. AINSWORTH, RUFUS C., Clarkson, N. Y.—Lieut. W. S. Camp, says:

"He was enlisted at Hamlin, N. Y., by J. E. Lee, October 21st, 1861, and was mustered into the United States service, at Buffalo, N. Y., by Lieut. Cutting, on the 26th October, 1861, to serve three years unless sooner discharged. November 1st, 1862, he was promoted to first duty sergeant by J. E. Lee, Captain Commanding. About September or October, 1863, he received a furlough of thirty days to go to New York, and did not return until he was arrested as a deserter, and returned to the command as such in February or March, 1864. On account of his desertion he was reduced to the ranks by Capt. L. A. Cady, January 20th, 1864. In April, just before the battle of Plymouth, he was sent, under guard, to New Berne, N. C., for trial by general court martial. The company all being captured (April 20th,) at Plymouth, before his trial, there remained no evidence of his guilt, and no one to appear against him. In this confinement he remained until, through the influence of Corporal Stoddard, myself, and the Christian Commission, his case was brought before the authorities, and he was released and returned to the company, then under command of Capt. E. De Meulen, at Roanoke Island, August 20th, 1864. During his confinement at New Berne, he was put in a gang of men, and worked in the Sanitary Gardens, so that he did not have a severe time.

"In accordance with Special Order No. 1, Head-Quarters Twenty-Fourth Independent Battery, Roanoke Island, August 20th, 1864, he was again promoted to sergeant, by Capt. E. De Meulen, commanding detachment, and held this position until November 7th, 1864, when, by reason of expiration of term of service, he was mustered out at New Berne, N. C.

"He is now farming in Wisconsin."

3. AINSWORTH, WM., Clarkson, N. Y.—Joined for service, October 8th, 1861. Re-enlisted as a veteran, in January, 1864, and was taken prisoner at Plymouth, N. C. He was a tall, strong, good-natured fellow, and made a capital No. 1 at the piece. Ferguson writes that Ainsworth died at Charleston, S. C.

4. ALBURTY, WILLIAM, Perry.—Joined for service, October 22d, 1861. He was but eighteen years of age on the muster roll, and it is our opinion that he was nearer sixteen than eighteen when mustered in at Buffalo.

He held the position of guidon in the Battery, and in drill, march or action, was prompt, ready and efficient. We quote the following from correspondence to the "Wyoming Times," August 15th, 1862:

"NEWPORT BARRACKS, July 31st, 1862.

"Last Friday we received orders to prepare for a march. We marched all day Saturday, and part of the night. Sunday morning we started again, marching about fifteen miles, when we stopped to feed the teams and eat our dinner. When we had been there about an hour and a half, we heard some guns fired and orders came to hitch up. This we did as soon as possible, and had hardly finished when a body of cavalry, three hundred strong, came charging right upon us. We wheeled our guns into position, and commenced firing. The fight lasted about an hour. Our force consisted of six companies of infantry, three companies of cavalry, and one section of our Battery. There were three or four of the infantry wounded; and two or three of the cavalry. There was no one hurt in our Company. Col. Heckman said, we worked our guns as well as any battery he ever saw. * * * * * William acted bravely. He rode up to where the Colonel was, and that was when the rebels were firing, and the bullets went by like hailstones. He had his flag in one hand and his revolver in the other. The Colonel told him to go back with his flag and horse and then come and if he got a shot to fire. The Colonel said he had good blood."

He re-enlisted in January, 1864; was taken prisoner at Plymouth, being one of the furloughed veterans who returned just in time to be captured. He died at Andersonville, Ga., August 23d, 1864.

The number of his grave is 6,698.

5. ALBURTY FRANCIS M., Perry.—Joined for duty October 21st, 1861. He writes:

"I enlisted in Perry, about the middle of September, 1861, and was mustered in at Buffalo, October 1st, 1861. I was mustered out at Plymouth, N. C., January 1st, 1864. Re-enlisted the same day as veteran, was absent at the battle of Plymouth, being delayed from returning with the other veterans by sickness, was transferred to Third New York Artillery, May 25th, 1865, was mustered out lastly at Syracuse, on the 7th of July, 1865. I was promoted to the rank of corporal, the 14th of April, 1863; was in the skirmish at the White Oak River; also in one at Kinston, and another at Goldsboro'. I am at present working at my trade, printing, in the "Silver Lake Sun" office, in Perry."

6. ALLEN Z.—Enlisted at Whitehall, and mustered in September, 1861. Re-enlisted January 1st, 1864.

Was married while on the furlough which had been granted him as a veteran. Zeph's white team, and his management of them on the lead, was one of the noticeable things of the Battery, during its days of drill. He was taken prisoner at Plymouth. Comrades will remember him as quite small, but hardy and tough. Prison life did not affect him to the extent that it did many of the larger and seemingly stronger men. He was paroled and exchanged in December, 1864; joined his company at Coanjock Bridge, N. C., April 23d, 1865. Was in good health, fat and fair, having been gone from the company one year and three days. Transferred to Third New York Artillery, May 25th, 1865; was mustered out in June, 1865.

His present address is Whitehall, N. Y.

7. ANDRUS, LEMUEL, Perry.—Joined for duty, October 25th, 1861. While the Rocket Battalion was stationed at Washington, he was seized by that terrible pestilence, the small pox, and died in the hospital, the 9th of March, 1862. We had not a personal acquaintance with him, but we always heard him spoken of by his comrades as one who highly deserved their friendship and respect.

8. ANDREWS, MARK.—Enlisted at Perry, N. Y., Oct. 21st, 1861. Mustered in at Albany, Oct. 26th, 1861. He was soon after promoted to position of sergeant, and afterwards to orderly sergeant. Through some differences and misunderstandings with the officers of the company, he was reduced to the ranks.

In January, 1863, he was mustered out to receive

promotion to first lieutenant in Tenth New York Artillery; in which he served until it was transferred to Heavy Artillery. He then resigned and accepted a commission as first lieutenant in the Twentieth New York Battery. On July 20th, 1864, he resigned this last commission, and soon after received an appointment in the Treasury Department at Washington. He must have proved a valuable man in the position he occupies, since he seems to hold it, notwithstanding change of Administration and of the Cabinet.

His present address is "Secretary's Office," Treasury, Washington, D. C.

Mark Andrews, Junr's, address is the same as his father's.

9. ARMSTRONG, WILLIAM.—Joined for service October 15th, 1861. Re-enlisted as a veteran, January, 1864. Was taken prisoner at Plymouth, and died at Andersonville, October 26th, 1864, of scorbutis.

10. ARMSTRONG, J. H., Mount Morris.—Mustered in, August 30th, 1862. Was discharged from the service, at hospital, for inability, some time in 1863.

11. ATWOOD, GEORGE S., Perry.—Was mustered in at Buffalo, August 30th, 1862. He died in the Andersonville Prison Stockade, August 28th, 1864, of chronic diarrhœa. The number of his grave in the cemetery is 7,207. The writer has always imagined that George had a feeling or presentiment that he would not return home. While we were at the Park Barracks, in New

York, he met his father. In a letter from Will Hosford to a friend I find the following statement of the affair:

"The first day that we were here some of the boys saw George Atwood's father, and told him George was here. They called George, and, without saluting him, the father began to abuse him for enlisting, told him he could not go, and said he should take measures to get him out. He finally came in and saw the lieutenant. But both Hastings and George remained firm. He went to the Mayor's office to get a writ of *habeas corpus*, but did not succeed. He told some of the boys that he hoped George would be shot, and would never return alive." A short time before the battle of Plymouth, George was troubled with hernia, and could have obtained his discharge from the service on that account, but when advised so to do, his idea of duty caused him to spurn the suggestion.

George was a kind-hearted, generous young man; unselfish and ambitious only to be well read, and able to meet any argument in politics or any of the ordinary topics of the day. In prison he maintained his character of consideration and kindness, and died beloved by all his comrades.

12. ANSBACHER, MOSES, Albion, N. Y.—Joined for duty Nov. 7th, 1861. Soon after the arrival of the new men at Newport Barracks, in 1862, Ansbacher obtained his discharge, in order to proceed to Germany, and take charge of some estate to which he had become heir. We have no further trace of him.

13. BAKER, JOHN, Covington.—Mustered in at Buffalo, August 30th, 1862. Was taken prisoner at Plymouth, and died in Andersonville Prison Stockade, Sept. 8th, 1864, of scorbutis. The number of his grave is 8,215. We do not remember having seen him but once after reaching Andersonville, and therefore do not know any particulars of his death.

14. BARKER, GUSTAVUS.—Enlisted at Clarkson. Mustered in at Buffalo, October 21st, 1861. Re-enlisted as a veteran in January, 1864.

Possessing a keen appreciation of the ridiculous, and an admirable adaptability in originating and carrying out schemes of fun, he often made the camp merry with some prank.

He was taken prisoner at Plymouth, was exchanged at Annapolis, and is now living at Clarkson Corners, N. Y.

15. BARNES, ROSWELL, Perry.—Was mustered in at Buffalo, August 30th, 1862. At the time we shipped on the steamer "Oriole" from New York to New Berne, Barnes and McCrink left the steamer for a few moments to make some small purchases. During their absence we cast off from the pier, and as they emerged from the grocery, they spied us out in the stream. They immediately jumped into a barge, and offered the oarsmen quite a sum of money as an incentive to overtake us, but they failed in the attempt, and turned back a good deal alarmed and chagrined.

It was afterwards rumored in Perry that they had deserted. We find in the Battery correspondence to the

"Wyoming Times" of November 7th, 1862, the following comment:

"Barnes and McCrink arrived at New Berne two days before our company, and going on to camp, informed the captain that we were coming. To have any suspicion of their loyalty would be treating them very unjustly."

A favorite in his detachment, jovial, witty and shrewd, was "Uncle Barney."

Barnes was taken prisoner at Plymouth, and died at Andersonville, Sept. 14th, 1864, of scorbutis. The number of his grave is 8,821. In a diary belonging to Barnes, and given by his sister to the writer, we find the following on the page for Sept. 14th: "Barnes died during last night. I think he died while sleeping. I slept at the side of him, and was surprised to find him dead when I awoke this morning.—NAPOLEON B. NEAL, Middletown, Conn."

16. BARTLEY, JOHN.—Was mustered in at Albany in September, 1861.

His native town was Palatine Bridge, N. Y. Bartley was quite a Tom Thumb in stature, good-natured and jolly, and devoted to his horses. He re-enlisted at Plymouth in January, 1864, and, while home on furlough, was married. Was taken prisoner at Plymouth, and died in the prison at Florence, S. C.

17. BARTLETT, HARTWELL, Perry.—Mustered in at Buffalo, August 30th, 1862. Was taken prisoner at Plymouth, and died at Andersonville Hospital of scorbutis, September 4th, 1864.

The number of his grave is is 7,877.

18. BACHELDER, B. FRANK, Perry.—Was mustered in at Buffalo, August 30th, 1862. Promoted to corporal April 14th, 1864. Taken prisoner at Plymouth, and died at Andersonville, of intermittent fever, July 16th, 1864. The number of his grave is 3,447.

Bachelder had suffered more or less from fever and ague while at Plymouth, and during his sickness at Andersonville was removed to the hospital stockade in time to have saved his life, had we been provided with proper shelter and sufficient medicine. He died in a congestive chill. We thought that he would recover, since, up to the day of his death, he was able to walk about, and had the appearance of being stronger than he proved to be. The loss of his companionship was felt by all of us who were left. In his camp life and in prison life his Christian character was manifest. He kept his Testament with him to his death, and the writer often found him perusing its contents or conversing with his tent mates upon the precious promises it contains. We cannot doubt that he entered those heavenly realms where suffering and sorrow are unknown.

19. BEERS, L. M., New York.—Enlisted November 6th, 1861. He was taken ill soon after the company reached New Berne, and died at the hospital, June 14th, 1862. He was buried on the following day in the graveyard on the east side of New Berne.

20. BIRDSALL, GEORGE, Tarrytown, N. Y.—Enlisted at New York, November 10th, 1861. Mustered in at Albany, December, 1861. He was appointed gunner

corporal, and in our target practice proved himself to be one of the best gunners in the Battery. He re-enlisted at Plymouth in January, 1864, and was promoted by Capt. Cady to sergeant, which he afterwards resigned, preferring to take charge of a piece as gunner. He was taken prisoner at Plymouth. While at Andersonville he was detailed in the hospital, and acted as nurse for the sick. He was exchanged in December, 1864, at Jacksonville. We are indebted to him for an account of the terrible sufferings that our poor fellows were obliged to undergo while an exchange was being arranged at this point. He was mustered out of service at New York City, May, 1865. He then returned to Tarrytown, and settled there in mercantile business. May 16th, 1867, he was married to Miss Jennie Van Tassel. Is now with the firm of T. & G. Birdsall, Tarrytown, N. Y.

At the time of his exchange he was sent, with others, from Andersonville, by railroad, to Albany, Ga.; from thence they were marched to Thomasville, a distance of fifty-five miles, in three days. After some parley and delay, they were counter-marched this entire distance. Again they were marched to Thomasville, thence to Baldwin, to what was termed "Halfway House;" and from there they were sent into our lines at Jacksonville. The torture and cruelty to which these poor famished and weakened men were subjected, by these unnecessary and forced marches, through heavy sands and fields filled with the prickly lollipot, tells an additional tale of the horrible inventions of their tormentors.

21. BILLINGHAM, IRA, Broad Albin, Fulton Co., New

York.—Mustered in at Albany, November 28th, 1861. The service proved too hard for his constitution, and he was finally discharged, at New Berne Hospital, for inability. He has since died at Broad Albin.

22. BLOOD, WILLIAM, Brook's Grove.—Mustered in at Buffalo, August 30th, 1862. He was one of the finest specimens of a soldier which the Battery could boast of. He was willing, and no coward. Was taken prisoner at Plymouth. The writer saw him but few times at Andersonville. He was reported to have died at Florence, S. C.

23. BLAKE, WILLIARD D., Gainsville.—Mustered in March 2d, 1864. He was one of the recruits who reached Plymouth but a short time before it was attacked; our acquaintance with him, therefore, was a brief one. He was taken prisoner at Plymouth. He was sent from the Andersonville Stockade to the hospital, having had a severe "sunstroke."

A peculiar phase of his sickness was his craving for fat. He would exchange anything for a piece of fat bacon. He was as well cared for as possibly could be, and his physical appearance was indicative of pretty good health. But he suddenly began to fail, and on the 9th of July, 1864, he died. The number of his grave is 3,231.

24. BOIES, E., Moscow.—Was mustered in at Buffalo, August 30th, 1862. After his arrival at New Berne he was troubled with heart disease, and was sent to the hospital.

He was there discharged for inability. We can find no further trace of him.

25. BROOKS, JOHN, Moscow.—Was mustered in at Buffalo, August 30th, 1862. Brooks was a quiet, faithful soldier, and did his duty without a murmur. We can find little trace of him after he was taken prisoner at Plymouth. He was at Andersonville, and stood the exposure and suffering at that prison better than the average. Ferguson reports that he finally died at Florence, S. C. Ferrin says that he died in October, 1864, at Florence, S. C.

26. BRAYTON, RUFUS, Perry.—Mustered in at Buffalo, August 30th, 1862. Brayton was the tallest and largest man on the muster roll, and when enlisted, was expected to endure more than any other recruit. He was known by his comrades as "Our Infant;" but from the time we left New York up to the time of his decease, he was afflicted with one malady or another, until he was so changed in appearance that he was hardly recognizable.

He died April 14th, 1863, at Plymouth, N. C.

In a private letter written by A. Lent, we find the following particulars: "Brayton had been but a short time with us, he was discharged from hospital at Fortress Monroe, and ordered to report at camp for duty, and while he was in camp at New Berne he did light duty, and appeared to be gaining strength.

"When the second section was ordered here, he came up with the boys, and soon after his arrival here he took cold, and had another touch of diarrhœa, and was sent to the hospital. I went to see him at the hospital, and he told me he had the billious fever. I did not see the surgeon then, but he told me a few days afterwards that Brayton was better. About 11 o'clock A. M., April 14th, word was sent up that he was dead. I learned that he died of pneumonia, and very suddenly."

It is rather singular that Lent died of pneumonia at Andersonville, a little over a year from the time of writing this letter.

27. Brown, George.—Enlisted October 18, 1864, at Rochester, for three years. Joined at Roanoke Island, December 16th, 1864. Transferred to Third New York Artillery.

28. Bullock, Robert.—Enlisted at Hamlin, October 21st, 1861; was mustered into the United States service by Lieut. Cutting, October 26th, 1861, at Buffalo. Was mustered out at Rochester, N. Y., July 13th, 1865, and did not re-enlist. Was a prisoner at Andersonville, Ga., also at Charleston and Florence, S. C. He was released at Charleston, S. C., on the 10th December, 1864, but not paroled at the time. When at Albany, N. Y., he was promoted to sergeant. Has occupied himself with farming, but during the past two years has been unable to work, in consequence of his impaired constitution—the effects of his protracted imprisonment. Present address, North Parma, Monroe Co.

29. Bulkley, W. E.—Joined for duty and enrolled for service at Castile, N. Y., February 15th, 1864.

Joined the Battery at Plymouth in time to participate in the battle, and be taken prisoner. April 1, 1864. He was reported on the roll as "Absent at College Green Barracks, as an exchanged prisoner," in October, 1864.

30. BULKLEY, CHARLES.—Enlisted at Castile, N. Y., February 15th, 1864. Joined the Battery April 1st, 1864. Was taken prisoner April 20th, 1864. We cannot find anything more about him, among all the information in our hands.

31. BUCK, ROBERT, Perry.—Joined for duty, October 3d, 1861.

Buck was engineer of the Ambulance. He was discharged from service, at New Berne, in May, 1862, for physical inability.

32. BUTTON, JAMES, Cuylerville.—Mustered in at Buffalo, August 30th, 1862. Was taken prisoner at Plymouth; and died at Andersonville Prison Stockade, of chronic diarrhœa. The number of his grave is 5,805.

33. BURD, HENRY C., Tarrytown.—Joined for duty, November 6th, 1861.

Mustered in December 1st, 1861. As a bugler, Burd had few superiors. His prompt responses to the commands of the drilling officer, were a great assistance to the excellence of the Battery movements, while on the drill ground. His musical talents and skill were often a source of pleasure to the members of the company.

He re-enlisted as a veteran, January, 1864; and while

on furlough, was taken sick, and by being thus obliged to remain home, was saved from the Plymouth capture. Was discharged at Roanoke Island, the 22d of November, 1864, by special order, War Department, No. 361.

At present he is employed at watch making; and we understand has become quite a musical artist.

His address is Tarrytown, N. Y.

34. CALHOUN, G. W., Albany, N. Y.—Enlisted for one year, September 5, 1864: Promoted corporal, December 10th, 1864.

Transferred to Third New York Artillery, May 25th, 1865.

35. CALTEAUX, PAUL, Perry.—Mustered in, August 30th, 1862.

He was appointed an artificer, and, as a rule, was about one of the busiest men in camp, as our horses must be shod, and our gun carriages and caissons must be repaired. His broken French-English jargon was either a terror or a sport to us; much depending on his humor, whenever we wanted work done. He was taken prisoner at Plymouth.

We lost track of him at Andersonville.

He was reported to have gone to work for the rebels, at his trade, in Charleston, S. C. Others affirm that he died at Andersonville.

36. CALKINS, JAMES, Perry.—Mustered in, August 30th, 1862.

Was taken prisoner at Plymouth, N. C. While at

Andersonville, he was sent to the hospital, and improved in health to that extent, that he saw he should be returned to the stockade. He thereupon, with others, planned an escape. The writer recollects occupying several days in obtaining and furnishing him with extra provisions and medicines, that he thought he would need for his subsistence, while on his uncertain tramp. Ferrin joined with him in the attempt to escape. They succeeded in getting away under cover of night; but in a day or two, were brought back into the prison camp. He was reported on his company roll as absent at College Green barracks.

Ferguson says he died at Florence, S. C.

He had little mercy for rebels, or neutral men, and in our marches, interpreted the Confiscation Act freely. Particularly if he thought there was any honey, chickens, eggs, ham, or other commodity that would make a good supper for his camp mess.

37. CAMP, WM. S., Perry.—He writes: Enlisted in Perry, N. Y., August 30th, 1862, and same day was mustered into service at Buffalo. November 1st, was promoted by Capt. J. E. Lee to sergeant. December 1, 1863, was promoted to quartermaster-sergeant by Capt. Cady. February 20th, 1864, with recommendations from commanding and company officers, made application to the Secretary of War for permission to appear in Washington, D. C., before Major-General Silas Casey's Board of Examination, for an examination as to fitness for receiving a commission in some light battery of artillery formed of colored troops. On the 12th of March, 1864, I received permission to appear in Washington, D. C.

I started, in company with Corporal S. A. Stoddard, for Washington, D. C., April 3d, 1864. Arrived there April 9th, and on the 11th appeared before the Board, but was informed that they were not granting or examining for commissions in artillery, and was ordered to return to the company. Applied to and received from Secretary of War a furlough of twenty days. Surrendered up my furlough, April 14th, and took an order to return to the company. April 15th, arrived at Norfolk, Va., where, missing the boat connection through the Chesapeake and Albemarle Canal, was obliged to wait over until Monday, 17th, when we (Stoddard and I) started for Roanoke, where we arrived the 18th. Had we made the connection, as usual, on the Saturday morning at Norfolk, we would have been in Plymouth Sunday evening, as we had calculated; but thanks for Divine interposition in our behalf, we were delayed. Heard first of the fighting at Plymouth when we arrived at Coanjock Bridge and changed boats for Roanoke. The steamer "Massasoit," which had left Plymouth late on Monday night with women and children, arrived at the island early on Tuesday morning, and as soon as she could be coaled up, started back for Plymouth. We returned with the "Massasoit," and when about two-thirds of the way up the Sound, we met a gunboat, having on board the remains of the gallant Capt. Flusser, and bearing the sad intelligence that the ram "Albemarle" had come down the river at two o'clock that morning, and had sunk the "Commodore Perry," and driven the "Miami" into the Sound, thus leaving the ram in full possession of the river. We steamed forward, and joined our fleet near the mouth of the Roanoke River.

During that day several refugees were picked up, who had escaped by coming down on the land side to nearly opposite the fishery, and below the ram, and then taking a dug-out, escaped. Learning that this passage to Plymouth was unobstructed by the enemy, Lieut. Langworthy, of the Eighty-fifth New York Volunteers, Stoddard and myself, tried to get a small boat to go up to Plymouth in, but without avail, luckily.

Wednesday afternoon, an old " darkie," who lived below the town, was picked up. He reported that the garrison had surrendered, and that he saw the flag on Fort Williams hauled down. Not having heard any firing of guns for several hours, and the arrival of other refugees corroborating the testimony of the old " darkie," we were convinced that the entire garrison of Plymouth had been captured by the enemy. Several " transports" arrived on Wednesday from New Berne with troops, but finding they could be of no service, they returned immediately. On board one of these boats I found Thomas McGuire, William Roach, Dennis Finnegan, Lawrence Green, Andrew J. Secor and Philemon Farrell; all but McGuire being recruits for the battery. These I had transferred to the " Massasoit," and I was ordered on board of a sutler's propeller with my squad, placed in charge of a load of refugees, and ordered to proceed to Roanoke Island and report to the commanding officer. Thursday, April 21st, 1864, arrived at Roanoke Island, and, with Corporal Stoddard and the six men before mentioned, reported to Lieut.-Col. Clark, Eighty-fifth

New York Volunteers, commanding at Roanoke Island, these were all of the Twenty-fourth New York Independent Battery, for duty. Corporal Stoddard was ordered to report with the men to Capt. Barnum, Sixteenth Connecticut Volunteers, commanding Fort Parke at the north end of the island. I was detailed to report to Capt. George C. Wetherbee, C. S. and A. A. Q. M., for duty in the Quartermaster's Department. I remained as clerk in the Quartermaster's office on Roanoke from April 21st, 1864, until February 6th, 1865. On the 28th of January, 1865, I received a commission as first lieutenant, to rank from December 28th, 1864, *vice* George S. Hastings, resigned. On the 23d of February I assumed command of the battery at Fort Foster, Roanoke Island. March 2d, was ordered with my command to Shallow Bag Bay, on the east side of the island, and on the 4th was ordered from headquarters, District of Beaufort, to proceed to Coanjock Bridge, on the Chesapeake and Albemarle Canal, and to assume command of that station. Remained in command of this station until May 25th, when, in accordance with Special Order 52, District of Beaufort, reported to Colonel C. H. Stewart, commanding Third New York Heavy Artillery, at New Berne, N. C. On the 26th of May, 1865, in accordance with Special Order No. 113, War Department, I transferred all the enlisted men of the Twenty-fourth Independent Battery over to the Third New York Artillery, and on the 29th of May, 1865, was mustered out of the United States service, at New Berne, N. C. Here ended my military career.

Present address, Lockport, N. Y.

38. CAMP, GEORGE, Mount Morris, N. Y.—Mustered in, April 1st, 1864.

Promoted corporal, October 21st, 1864.

Transferred to Third New York Artillery, May 25th, 1865.

39. CANFIELD, S. D., Syracuse, N. Y.—Mustered in, September 20th, 1862. He was one of the students of Hamilton College who joined our ranks. His stay with the Battery was short. He was promoted to a commission in a New York Cavalry Regiment. We have since heard of his making fortunes in oil at the time of the oil fever.

We cannot ascertain his present address.

40. CARNAHAN, WM., Moscow, N. Y.—Mustered in, August 30, 1862.

Taken prisoner at Plymouth, N. C.

While at Andersonville, he was sent to the hospital, and recovered so as to be able to assist in nursing the sick.

At the time of the rumored exchange, he was sent with one of the first squads that were supposed to be going to Savannah. But the writer, on arriving at Millen, was surprised to find Carnahan there, feasting on sweet potatoes, and seemingly in improved physical condition.

He was soon after exchanged, and was finally mustered out, at Syracuse, July 7th, 1865. We afterwards heard of him, as a "Canvasser" for some of the works on "Prison Life."

He was married after leaving the army. His present address is Hubbardston, Michigan.

We give the following in his own words, under date of May 9, 1869:

"I was taken prisoner at Plymouth, N. C., 20th April, 1863. Was taken to Macon Prison, Ga. Was there a few days, and then sent to Andersonville. I remained there five months. I was in the stockade two months, and then I was taken with the typhoid fever, and sent out in the hospital. I had been there three weeks before I knew where I was. After I came to myself, I found that *   *   *   had taken care of me, and fed me. I consider I owe my life to him.

"From the village of Moscow, there were twenty-two enlisted in this company, out of which returned two—Andrew Ferguson and myself. They all died in prison, but two. Murray Grant died at Plymouth, of sickness, and George Meade was shot in the battle when we were captured. I saw him after he was dead. The rest were taken to Andersonville, to their long home. The only brother I had was with me. I parted with him the 10th of September. The poor fellow lies outside the Andersonville stockade. He starved to death. At the time he died, I was not able to walk; so some of my comrades carried me in a blanket, to bid him good bye for the last time.

"I was taken out of Andersonville, after being there five months, and sent to Savannah, where I remained three weeks—and a long three weeks it was. From there I was sent to Millen, and there remained until the rebels routed us back to Savannah. We left Millen in the morning, and Sherman's cavalry came there at night. If we had been there twelve hours longer we would have been captured back. So after we got to Savannah, the rebels made up their minds they had better let us go. So after we had been there three days, they paroled us, and put us on board of a steamer, and sent us to the mouth of the Savannah river, where we met one of the Yankee boats. It was the happiest day that I ever saw. Paroled, November 25th, 1864. After I was paroled, I was sent to Parole Camp, Annapolis, Md., and then got a furlough of sixty days. When I got home I weighed eighty pounds, and during the sixty days of my furlough I gained a pound a day. After my furlough, I returned to Annapolis, and there was on detail, as orderly, for Col. Hutchins, who had charge of issuing checks for prisoners commutation ration money at College Green barracks, Annapolis. I was there three months, and then sent to my company, to New Berne, N. C. When I got there I found about twelve of the old boys. It seemed terribly lonesome to find so few of my old comrades, and to know how they were treated, and died—starved to death—and a near and dear one to me had gone with them; the only brother I had in the world."

41. CARNAHAN, CHAS., Moscow.—Joined for service, December 19th, 1863. Joined the company for duty, April 1, 1864.

He was one of the recruits who arrived at Plymouth just in time to participate in the battle, and be taken prisoner. He died at Andersonville Stockade, of scorbutis, September 11th, 1864. The number of his grave is 8,470.

42. CHAPMAN, JOHN, Perry.—Mustered in at Buffalo, October 26th, 1861.

Chapman was the wagoner, and in the post he occupied, he had both opportunity and disposition to domineer in his particular province. On several scouts or foraging expeditions, the boys retaliated a little, by starting a scare of "rebels approaching"! which brought out from John a wonderful display of cracking a whip, and handling four-in-hand, on a galop. We recall the time when Sergeant Camp was fired at by our own vidette, and an alarm raised in camp, which brought out a squad under Captain Cady, to meet and assist us. As soon as John fully comprehended that rebels were reported near, there was a blanched face—a gathering and tightening of the reins—a goad of a long lash, and a yelp to his team, that would have fairly awakened a Rip Van Winkle. We finally overtook him, and had our laugh at him. He did not care to hear the story afterwards.

He re-enlisted at Plymouth, in January, 1864, and received a veteran furlough. He was never heard of afterwards, and was reported as a deserter.

43. CHADBOURNE, HENRY, China, N. Y.—Mustered in, August 30th, 1862.

Was taken prisoner at Plymouth. Being of a rather frail constitution, he was one of the early ones that was grasped by that terrible disease, chronic diarrhœa, and for want of proper nourishment and medicines, he rapidly run down until he died. He died June 18th, 1864.

The number of his grave is 2,157.

44. CHAPIN, WM. E., China, N. Y.—Mustered in, August 30th, 1862, at Buffalo.

Shortly after his joining the Battery, he was ill, and was sent to the hospital. During his convalescence, he proved to be a good nurse, and continued in the hospital in different positions. Was connected with the Hospitals and Sanitary Garden all the time he was in the service. He was finally mustered out of service, and returned to his home—China, Wyoming Co.

Present address, La Salle, Ill.

45. CLARK, C. A.—Enlisted, October 1, 1861. Mustered into service, October 26th. Promoted to quartermaster sergeant, in June, 1862. Assigned the command of a detachment, as duty sergeant, at Newport barracks, in December, 1862. Re-enlisted, as a veteran volunteer, at Plymouth, in February, 1864. Went north, on leave, in March. Received second lieutenant's commission in Sixteenth New York Artillery, but on account of sickness was unable to accept it. Remained in Perry, N.Y., until January, 1865. Commissioned as second lieutenant in Twentieth Battery, on duty at Governor's Island, N. Y. H. Battery removed to Battery barracks, N. Y. City, in May. Promoted to first lieutenant, in May, 1865. Mustered out, August 5, 1865.

Entered the employ of the United States Telegraph Co.; remained until its consolidation with Western Union Telegraph Co. Then entered the office of the "North America" Life Insurance Co., where he continues.

Present address,—care of North America Life Insurance Co., 17 & 19 Warren Street, New York City.

46. CLUTE, H. V., Cuylerville.—Mustered in, August 30th, 1862.

He acted as assistant artificer, with Calteaux.

Was taken prisoner at Plymouth. He was not in good health when captured, having suffered from chills and fever while at Plymouth; and this disease seemed to cling to him after he reached Andersonville. He died, May 30th, 1864, of intermittent fever. The number of his grave is 1,497.

47. COMSTOCK, A. W., Perry.—Mustered in, August 30th, 1862.

During the battle of Plymouth, he had charge of the horses attached to the limber chest of Crooker's piece, which was posted at the parapet off from the right of our park at Plymouth. The rebels having taken the little fort occupied by Capt. Chapin, of the Eighty-fifth New York, had concentrated their fire on the embrazure which Crooker's piece occupied. One of the rebel shells or balls passed through the embrazure, struck the limber chest, and caused an explosion of the ammunition which it contained. By this explosion Comstock was wounded, and some of the horses which he was in charge of were killed and some wounded.

Comstock was carried to the hospital, said to be wounded in the thigh. He died May 9th, 1864.

48. COOK, HARLO, Hamlin.—Joined for service, October 26th, 1861.

Discharged at Academy Hospital, New Berne, for inability on account of dropsical affections, June, 1862.

49. CORBIN, B. F., Hamlin.—Enlisted by Lieut. Cady, and mustered in at Buffalo, October 25th, 1861. He was appointed corporal, January 1st, 1864, and promoted to sergeant, April 14th, 1864. He re-enlisted in January, 1864; was taken prisoner at Plymouth, and died at Andersonville, June 9th, 1864, of chronic diarrhœa. The number of his grave is 1,778.

He was among the first of our comrades who fell before that terrible disease. Being a strong, hearty man, it was impossible for him to supply the demands of his appetite by means of the usual prison rations. Want of food was the beginning of his illness, and the corn bread which was furnished him only tended to irritate his stomach and aggravate the disease. We were finally enabled to purchase some milk and some berries for him while in the hospital, but it was too late. He was conscious that death was near, and was perfectly resigned. He died easily and almost imperceptibly.

50. CORKWELL, JOHN, Rochester, N. Y.—Enlisted for one year, September 27th, 1864. Transferred to Third New York Artillery, May 25th, 1865.

51. COWEN, JAMES, Albany.—Mustered in at New Berne, May 21st, 1862. He was absent on furlough at the time of the capture of Plymouth. Was promoted to corporal, December 10th, 1864. Mustered out at Albany, in June. Present address, Albany, N. Y.

52. CROOKER, WM. W., Perry.—Joined for duty, October 3d, 1861. Re-enlisted as a veteran, January 1st, 1864.

Promoted to orderly sergeant, April 14th, 1864. Taken prisoner, April 20th, 1864. Appointed by Governor Fenton of New York as captain of the battery, *vice* L. A. Cady, resigned, to date, January 10th, 1865. Transferred to Third New York Artillery as orderly sergeant. Crooker had charge of one of the divisions in the Andersonville Hospital, and did all in his power to alleviate the sufferings of his fellows. He still bitterly resents the treatment of the men at that stockade, and theorizes that our Government was as much to blame as was the Confederate Government. After his discharge from the army he visited the Southern States. He finally married there, and has settled at Jeffersonville, Indiana. We are sorry that his business duties were so overwhelming that he could not spare the time to write us a sketch of his experience.

His travels in the process of exchange were in quite a different direction from that which most of us were obliged to participate in. We understand that Crooker was in the same squad with Birdsall. (See Birdsall's personal sketch.)

53. Crooks, J.—His name appears on the roll, but we know nothing of him.

54. Crosby, M., Sardinia, N. Y.—Mustered in at Buffalo, October 26th, 1861. Re-enlisted at Plymouth in January, 1864. Was taken prisoner at Plymouth and died at Andersonville Hospital, of typhoid fever, July 14th, 1864. The number of his grave is 3,324.

He was an athletic man, a good soldier, a whole-souled fellow. He endured everything cheerfully and bravely.

55. CROUNCE, GEORGE, Albany.—Was mustered in at Albany, in September or October, 1861. His peculiarly feminine appearance gave him the sobriquet of "Miss Crounce." His tastes and pursuits were as girlish as his form. His tent was always tidy and in order and his culinary skill unsurpassable. He was too a good man at his post, at the piece, and showed coolness and bravery in battle.

He re-enlisted at Plymouth, in January, 1864. Was taken prisoner and died at Andersonville Stockade, of intermittent fever, June 20th, 1864. The number of his grave is 2,273.

56. CUSICK, HIRAM, Rochester, N. Y.—Enlisted for one year, October 10th, 1864. Transferred to Third N. Y. Artillery, May 25th, 1865. Joined, November 9th, 1864.

57. CULVER, A. L., Gainsville.—Mustered in, August 30th, 1862. Was taken prisoner at Plymouth; he had been suffering from intermittent fever in Plymouth, and the experiences of Andersonville did not aid him in recuperating. He died in the stockade, of chronic diarrhœa, July 27th, 1864. The number of his grave is 4,119.

58. CYPHER, GEORGE W., Kingston, N. Y.—Enlisted for one year, September 9th, 1864. Transferred to Third N. Y. Artillery, May 25th, 1865. Joined October 18th, 1864, at Roanoke.

59. DAVIS, ORNAU, Leicester.—Mustered in, January 5th, 1864. Arrived at Plymouth in time to take part in the battle and be taken prisoner. Joined for duty, April 1st, 1864.
Ferguson reports that he died at Charleston.

60. DOLBEER, CHAS. H., Perry, N. Y.—Enlisted on the 25th of August, 1862, at Perry, Wyoming County, N.Y., and was mustered in August 30th, at Buffalo. Joined the Battery, the 25th September, at Newport Barracks, N. C.

On the 4th of November was appointed orderly-sergeant. January 6th, 1863, by orders from Brigade Head Quarters, General Ledlie commanding, was promoted to junior second lieutenant. January 13th, 1863, by orders from same head quarters, (General Ledlie,) was detailed as "Chief of Brigade Ambulance Corps," and reported to Chief of Division, Ambulance Corps, for duty. March 17th, relieved from duty on Ambulance Corps. Was commissioned as second lieutenant, June 23d, 1863—(See Adjutant-General's Report, N. Y. State, 1868.) Was mustered out at Plymouth, N. C., January 21st, 1864, in compliance with orders from War Department, dated January 6th, 1864. Discharged from service, January 22d, 1868.—(See report above referred to.) While on detached service in Ambulance Corps, January 26th, 1863, in compliance with orders, reported on board General Foster's flag ship, "Spalding," off Morehead City; arrived off Charleston, January 31st, and at Port Royal, February 2d. Returned to New Berne, February 11th, 1863. While on the same service, March 7th, 8th, 9th, and 10th, went with the expedition up the Trent. On the 8th, found the Rebs at White Oak River. Having been relieved from this duty, moved, March 26th, with the centre section of Battery, on board steamer "Escort," lying at New Berne, and arrived at Plymouth, the 28th. July 26th, moved with centre section to Jamestown. In skirmish at Foster's Hill, the 27th, and returned to Plymouth, the 29th.

Having obtained leave of absence to go out of the Department (which extended to Fortress Monroe), for twenty days, left Plymouth, on the 17th September, 1863, and visited friends in New York State and in Michigan. Returned to the command, the 17th October, 1863. The report to corps headquarters, in compliance with order No. 2, of the 19th August, 1863, not explaining the overstay to the satisfaction of the commission appointed to examine and report upon such matters, charges were preferred for "absence without leave"; and before a court martial, of which Lieutenant George S. Hastings, of the Battery, was Judge-Advocate, was tried and acquitted, as the following copy of general order, No. 51, will show.

HEAD-QUARTERS,
ARMY AND DISTRICT OF NORTH CAROLINA,
New Berne, N. C., Dec. 31, 1863.

General Orders,
No. 51.

Before a General Court Martial, which convened at Plymouth, N. C., on the 17th day of December, 1863,

pursuant to General Orders, No. 45, dated Head-Quarters, Army and District of North Carolina, New Berne, N. C., December 14th, 1863, of which Col. T. F. Lehman, One Hundred and Third Pennsylvania Volunteers, is President, were arraigned and tried : * * *
No. 2.
Second Lieutenant C. H. Dolbeer, Twenty-fourth Independent Battery, New York Volunteers :—Charge, " Absence without leave."

Specifications—" In this, that he, Second Lieutenant C. H. Dolbeer, Twenty-fourth Independent Battery, N. Y. V., having obtained leave of absence for twenty days —Special Orders, No. 52, Head-Quarters, Department of Virginia and North Carolina—did, on or about the 18th day of September, 1863, leave the camp of the Twenty-fourth Independent Battery, N. Y. V., near Plymouth, N. C., and did not rejoin his Battery until the 17th of October, 1863. All this near Plymouth, N. C.

To which charge and specification, the accused pleaded as follows :—" Not Guilty."

The Court, after mature deliberation upon the evidence adduced, find the accused as follows :
    Of the Specifications.. " Guilty."
    Of the Charge......... " Not Guilty."

The Court are of the opinion, that prior to circular, dated Head-Quarters, Army and District of North Carolina, New Berne, N. C., October 9th, 1863, that leaves of absence were generally construed to commence from the date of leaving the Department, and that construction is applied to this case. The accused reported in time, at Fort Monroe, and returned to his post as speedily as transportation was afforded. His leave of absence apparently allowed him twenty days from Fort Monroe.

The Court do therefore acquit him, Second Lieutenant C. H. Dolbeer, of Twenty-fourth Independent Battery, N. Y. V.

The proceedings and findings in the cases of *   * and Second Lieutenant C. H. Dolbeer, Twenty-fourth Independent Battery, New York Volunteers, are approved and confirmed.

By command of Major General PECK.

BENJ. B. FOSTER,
Asst. Adjt. General.

(Official.)

After returning from the army, resumed former occupation, as book keeper, in the office of Alva, Smith & Son, Batavia, N. Y., remaining one year. He then resided in St. Louis, Mo., being employed in the office of L. & D. Babcock, in the examination of land titles.

Present address,—Batavia, N. Y.

61. DURYEA, GEORGE, Perry.—Mustered in, November, 25th, 1861.

Deserted some time in April or May, 1862.

He afterwards returned to the company and did his duty as a soldier and like a man; was taken prisoner at Plymouth; exchanged at Charleston, December 6th, 1864. For a long time he was obliged to remain at home, on furlough, being dangerously ill, from the effects of his prison life. We have not been able to find his address, and therefore cannot give date of his discharge from service.

62. DURYEA, JOSEPH, Perry.—Enlisted with Samuel Andrus.

He deserted the company at the same time that his brother George and Grisewood did, and went to Canada. His name was dropped from the company roll. His present address—Elizabethport, New Jersey.

63. EASTWOOD, EDWIN M., Brooks Grove.—Mustered in, August 30th, 1862.

Eastwood was a cautious, steady soldier. A Christian, and a man who desired to do right with all whom he came in contact with. He was taken prisoner at Plymouth. He died at the Andersonville Hospital, July 30th, 1864, of chronic diarrhœa. The number of his grave is 4,410.

He was aware that death was near, and was prepared for it. He made a disposition of the few mementos he had, and we had the pleasure of returning them to his parents. In reply we received from them one of the kindest and most satisfactory communications that has been addressed to us since our return from the army.

64. FARRELL, PHILEMON, Rochester, N. Y.—Enlisted March 24th, 1864.

Transferred to Third New York Artillery, May 26th, 1865.

Found at mouth of Roanoke river, April 20, 1864, just after the Battery were captured. A lucky boy.

65. FERRIN, J. T.—We give his letter *verbatim et seriatim:*

"HOLLAND, N. Y.

"I enlisted the 18th of October, 1861, at Perry, Wyoming County, N. Y. Mustered in at Buffalo, October 26th, 1861. Mustered out, the 18th of July ,1865, at Syracuse, N. Y. I re-enlisted as a veteran, January 1st, 1864, at Plymouth, N. C. Was in the battle of Plymouth, the 18th, 19th and 20th of April, 1864, where I was taken prisoner.

"I was in Andersonville, Charleston and Florence prisons. I got to Andersonville the 1st of May, and there I had an introduction to old WIRZ; he put us into the stockade to live on one pint of meal a day and a teaspoonfull of salt every other day. It was rather tough fare at first. I was not quite as cunning as some of the boys when I was captured, for I did not take away blankets, for I supposed I was going to be taken better care of than I was. But when I got to prison I found out that I had got to have something to keep me warm and to keep the sun off, so I bought a blanket, and paid thirty-five dollars Confederate money. The next night, Morton Crosby, one of my tent mates, had his blanket stolen off him; so three of us boys clubbed together and bought another one. The 2d of June I went into the hospital. I was put into Dr. Barrow's ward; he did all he could for me, and cured me in a short time. By the 1st of July I was able to go around the stockade. I was there until the 9th of July, when James Calkins, of the Twenty-fourth, and John Burgoin, of the Eighty-eighth Illinois Regiment, thought that we would go home. So the night of the 9th of July we got over the stockade, down by the sink. When we got over, we went into the water almost waist deep. I was some time getting over, for when I got on the top of the stockade, the guard turned and came athwart me, and stood still for some time, I thought, but I finally got over all right. I had not gone far, when I ran foul of some brush, and in putting my hand out, I got hold of a snake. I suppose it was one; at any rate, it slipped through my fingers like one. We traveled around until morning, when we got out of the swamp, and laid in the woods all day. We traveled the whole of the next night again, and got along very well until the morning of the 12th, when we were near the swamp. We heard the dogs bark, and we made for some trees, but I could not climb, I was so weak. The dogs came up, but the men were right behind them, so they did not bite me. If it had not been for that, I should have been nibbled some. We were within about ten miles of Macon, having traveled about fifty miles ; in another week we would have been inside of our lines.

"They were old farmers that captured us, and had three blood hounds and a bull terrier. We went back to Andersonville, and old Wirz told us that we would be the last ones paroled, and he put me in the One Hundred and Sixth (106th) Detachment.

"I went inside of the stockade in time to see the raiders hung, and I stayed there until the 11th of September, when I went out with the Twenty-sixth Detachment, with the promise of going home. But we went to Charleston, and from there to Florence. By the time I got there I could not sit up, so I was put out of the cars by the side of the railroad, with about thirty-five others. We laid there until Dr. Dargin, a Rebel doctor, came to our relief. He put us into an old barn, and rigged it up for a hospital. He took good care of us. There was myself, John Brooks and Harry H. Foster, of the Twenty-fourth. There John Brooks died ; there Foster and myself lived. By the end of October I was able to go around, when one morning the doctor came down and said he had orders for all that were well enough to go to the stockade, and that made me sick again right away. But as soon as he told us that he had got a parole for all that would work for him, I felt better. And so nine of us worked from that time until the 8th of December, when we were paroled for good. I went from Florence to Charleston, and they put us into the Roper Hospital, where we stayed three days. The guard told us that we were to go back again to prison, which made us feel very bad, but the third morning we got on board the steamer ' Clemb,' and started for our boats near Fort Sumter. When we came in sight of the old flag, those who had caps took them off and gave one shout ; it was a glad one. But the Rebs shut us up with, ' You will go back to Charleston if you don't stop that.' I went to Annapolis, and there got a furlough home, and returned to Camp Parole, March 27th, 1865. Went from Camp Parole to Alexandria, Va., and from there to Norfolk, Va., where we took the boat through the Dismal Swamp Canal for North Carolina. Arrived at Coinjock Station, where my company were at camp, under Lieutenant Camp. Left Coinjock the 17th of

May. Arrived at Roanoke during the night; started for New Berne, on board the 'Ella May,' on the 18th. Awoke the next morning and found myself on board the boat 'Tay,' at Foster's Dock. We went into camp across the tent, in the same place that we did three years ago. On the 23d of June we got orders to get ready to go home. Went to Beaufort, and started on board the 'Edward Ewett' for New York; thence to Albany, and from there to Syracuse, where we were discharged the 18th of July, 1865. Married, November 2d, 1865, to Helen M. Cheney, of Holland, Erie County. Present residence, Holland, Erie County, N. Y. Occupation, blacksmith."

66. FERGUSON, ANDREW T.—Enlisted August 30th, 1862, at Moscow.

Mustered in at Buffalo, Sept. 10th, 1862.

Received a corporal's warrant.

Ferguson was a good gunner and made some capital target shots. Was taken prisoner at Plymouth, N. C. Ferguson's music, which was always such a pleasure to the Battery boys, charmed even those Southern beasts, and an occasional desire among the Reb officers to hear a tune, put him in favor with them. It was during one of these respites that he witnessed the destruction of the letters, as told elsewhere.

He was sent from the stockade to the hospital to do police duty, and from there went, on September 9th, 1864, with one of the first squads that were said to be going to Savannah for exchange. He, however, found himself shipped into the prison at Charleston. From there he was sent to Florence, and was exchanged at Charleston, December 6th, 1864.

After receiving furlough at Annapolis, he went home, and was very sick three months. He again reported to his company, and was mustered out at Syracuse on July

18th, 1865. He then resumed his former occupation of teaching dancing and deportment. We quote the following from "The Wyoming Sun," November 20th, 1868 :

"Prof. Ferguson proposes to open a school for dancing and deportment in this village as soon as the necessary arrangements can be completed. The successful and creditable manner in which his school was conducted last winter has given him a very favorable reputation as an accomplished teacher and a worthy gentleman, and we have no doubt he will meet with a liberal encouragement."

His present address is Cuylerville, N. Y.

The following is a copy of a letter received in reply to a query of ours :

"Yours of the 18th inst. came to hand in due time, making some inquiries concerning the dark days. It is still fearful to think of, concerning Wirz and some letters. About the 1st of September I went to his head-quarters tent to get a pass to go and play for a dance at his house, three miles away. At his tent were his wife's daughters, examining letters, reading every one, and destroying such ones as commented on the situation. (This a boy told me that lived with them.) The pile of letters I myself saw. After I had played for the party, I went to live with Dr. White, surgeon of the post. Then I saw the destruction of pretty much all of the letters that had been written by the prisoners up to that time. (This occurred on the evening that Stoneman shelled Macon, about the 6th of September.) The doctor brought out of the house more than two bushels of letters. We seated ourselves around the pile, and tore off the stamps, and opened as many as we liked. I saved and brought home one hundred and twenty-five stamps, and destroyed twice as many. 'We,' including the doctor, his mistress (she that claimed to be Provost-Marshal Reede's wife), myself, and an Irishman that was then making shoes. The letters made a pile as high as a table. This was about ten o'clock in the evening. He said they 'would drive away musquitos,' and put fire to them, and the wail of the poor prisoners ascended higher than the smoke.

Truly yours,

A. T. FERGUSON.

67. FILBIN, JOHN, Perry.—Mustered in, August 30th, 1862. Notwithstanding that, when at home, he was a strong man and a hard worker, change of scene and climate threw him into a chronic state of sickness. He was taken prisoner at Plymouth and is reported to have died at Florence.

68. FINNIGAN, DENNIS, Warsaw, N.Y.--Enlisted March 24th, 1864. Transferred to Third N. Y. Artillery, May 25th, 1865. Joined April 20th, 1864.
A quick witted lad.

69. FITCH, CHAS. W., Perry.—Mustered in, August 30th, 1862. Charley was a pleasant, sociable comrade. He understood and did his duty, and, therefore, made a faithful, commendable soldier. He was taken prisoner at Plymouth. He died at Andersonville Hospital, of pneumonia, August 4th, 1864. The number of his grave is 4,819.

70. FITZGERALD, THOMAS, Perry.—Mustered in, August 30th, 1862. Was taken prisoner at Plymouth and died at Andersonville Stockade, August 21st, 1864. We did not see so much of him after we reached Andersonville, and can, therefore, say but little about his sickness or death.

71. FITZPATRICK, PIERCE, Albany.—Joined for service, November 16th, 1861. Re-enlisted in January, 1864, as a veteran.

Fitzpatrick was a very peculiar fellow, a great speculator, a kind of battery sutler. His eagerness to "make a strike," when we were out on a march or a raid, often led him into trouble—(See account of the Trenton march). He was of a fierce and nervous disposition—even slightly bordering on insanity. Was always writing letters to the President, suggesting the best means of conducting the war. Writing so bad that no one but himself could read it, and he not when it had got cool. At the first of the attack on Plymouth, he caused considerable merriment among the members of the Battery by pursuing a portion of an exploded rebel shell, which had just passed by his head, with a velocity which, probably would not have been abated much had his head interfered. He considered that piece of shell very valuable as a "relic," and stowed it away in his pocket with the remark that "money couldn't buy it."

On the morning of the last day of the battle, he was stationed at the corner of intersecting streets, on the extreme left of the line of works, in charge of the caisson and caisson horses, belonging to Merrill's detachment. The rebels had broken through the left of the works, and were marching directly upon the right. As they approached, they called upon Fitzpatrick to surrender, but he unflinchingly refused. At this time the rebels fired a volley, simultaneously with a double-shotted canister discharge from the piece to which Fitzpatrick belonged,—and he fell. Nearly all of the Battery horses and many of the rebels were killed by the canister. No one can tell whether he was killed by the canister or by the bullets of the rebels,

72. FLYNN, JAMES, Hamlin.—Joined for duty October 21st, 1861. Re-enlisted, as a veteran, January 8th, 1864. He was promoted to corporal while at Plymouth. He suffered a great deal from chills and fever at Plymouth; at one time he fell down a flight of stairs in a congestive chill and was taken up for dead. Corporal Hurlburt was near with some restorative, and doubtless, by his prompt action, Flynn's life was saved. He was taken prisoner at Plymouth and died at Andersonville Stockade, September 10th, 1864, of scorbutis. The number of his grave is 8,378.

73. FOSTER, HENRY (*alias* Henry Frost).—Enlisted at Brooklyn, N. Y., on the 24th February, 1864. Mustered in at Riker's Island, N. Y., March 10th, 1864, and joined for duty at Plymouth, N. C., March 14th, 1864, where he was taken prisoner on the 20th of April, and was marched one hundred miles, to Tarboro, N. C., *en route* for Andersonville Prison. Remained there five months; was then transferred to Florence, S. C., and was in prison there three months. Was paroled at Charleston Harbor, S. C., on the steamer "Verona." December 11th, 1864, arrived at Camp Parole, Annapolis, Md. December 15th, got a furlough from 24th December, to January 25th, 1865. On the way homeward, was seized with typhoid fever, and taken to Camden Street Hospital, Baltimore, December 29th, where he was sick five months. He was discharged from service, May 13th, 1866. Married a young lady of Philadelphia, July 25th, 1865, at New Berlin, Union Co., Penn.

This young man saw an advertisement in the New York papers—" Wanted a Pay Master's Clerk," and as he was subject to draft, the best thing he could do would be to accept some such position and get out of it. He accordingly went to New York, and applied for the position. Was greeted cordially, asked to drink, and told that that position had just been filled, but that they had another position equally as good, that of captain's clerk. Drank several times, and then concluded to accept the position of captain's clerk, for the Twenty-fourth Independent Battery. The papers were drawn up, and he was told that before accepting the position it would be necessary for him to take the oath of allegiance to the Government, this he did, then drank the health of all hands, and knew no more until the next morning, when he woke up, dressed in uniform on Riker's Island. Was forwarded to the company and duly presented himself to Capt. Cady, March 10th, 1864, as his clerk, and was astonished and disappointed to find that he was a soldier, and was assigned to Camp's detachment as an extra duty man and had charge of the *Big Gray Horse.*

This was his story after arrival at the company.

Present address, No. 333 Franklin Street, Baltimore, Md.

74. GALUSHA, JONAS E., Perry.—Enlisted August 29th, 1862.

Mustered in, August 30th, 1862, at Buffalo.

Taken prisoner at Plymouth. Was exchanged, and died at the Parole Camp, at Annapolis, Md., December 19th, 1864, of chronic diarrhœa.

He was a quick, active soldier; and, we believe, was promoted to corporal, while stationed at Plymouth.

75. GOODHUE, D. W., Rochester, N. Y.—Enlisted and mustered in, September 30th, 1864, for one year.
Transferred to Third New York Artillery, May 25th, 1865.
Joined as a recruit, November 9th, 1864, at Roanoke Island.

76. GOULD, WILLARD, Moscow, N. Y.—Mustered in, August 30th, 1862.
Discharged, by reason of disability, by order of Major General Dix, August 20th, 1864, at General Hospital, Newark, N. J.
Think he was transferred to the Invalid Corps.
We have not been able to find his present address.

77. GRANT, MURRAY.—Enlisted at Moscow, August 30th, 1862.
Was mustered in at Buffalo, September 10th, 1862.
Promoted to artificer. The southern climate did not agree with him, and he was ailing for a long time.
He finally died at Plymouth. We believe his remains were embalmed and sent home.

78. GREEN, LAWRENCE, Moscow.—Enlisted for three years, January, 4th, 1864.
Mustered in at Canandaigua, January 23d, 1864.
Transferred to Third New York Artillery, May 25th, 1865.

Joined April 20th, 1864, at mouth of Roanoke river, en route to Plymouth. A good soldier.

79. GRIFFITH, CHAS. R., Perry, N. Y.—Joined for service, October 1st, 1861.

Re-enlisted, as veteran, January 1, 1864.

Was appointed corporal at organization of the Rocket Battalion, at Albany. Was promoted to sergeant, November 4th, 1862.

Taken prisoner at Plymouth, and died at Florence, S. C. He seemed to have little hope after his capture, and gave himself up to the belief that he should never return home alive.

80. GRIFFITH, ALBERT, Perry.—Enlisted August 27th, 1862.

Mustered in at Buffalo, August 30th, 1862. Was appointed artificer, November 4th, 1862.

He was the oldest of the three brothers. Willis had enlisted first, Charles had followed, and Albert felt that he could not stay at home. Therefore, when the enthusiasm of the enlisted men, in the fall of 1862, was at its height, he too was drawn into the vortex, and joined the common cause with them. While we were erecting our sheds and barracks in New Berne, his practical knowledge of building was invaluable to us. He was taken prisoner at Plymouth, and died in the Andersonville Stockade, of chronic diarrhœa, July 9th, 1864.

The number of his grave is 3,101.

81. GRISEWOOD, THOMAS, Perry.—Enlisted Nov. 22d,

1861. In company with the Duryea brothers, he deserted from the company while it was at Washington; and we believe he never returned to it. We are told that he is in California.

82. HART, CHAS.—Joined as a recruit Oct. 12th, 1864, at Roanoke Island. No muster and descriptive roll received. About fifty years of age. Said he was drugged and taken from New York City to Hart Island in August, 1864. Did not know whether he volunteered or was a substitute. Never received any bounty. Was a man about like Geo. McEwen, only faithful and orderly.

83. HARMON, JOHN C., Rochester, N. Y.—Enlisted and mustered in for one year, September 30th, 1864, and joined November 9th, 1864, at Roanoke. Transferred to Third New York Artillery, May 25th, 1865.

84. HARRINGTON, M.—Joined for duty November 9th, 1861, at Clarkson.

85. HASTINGS, FRED'K E., Mt. Morris, N. Y.—Was among the earliest recruits of the Battery. Serving as a private for many months, he afterwards received promotion to the several positions of sergeant, second and first lieutenant.

He participated in the battles attending the first Goldsboro' expedition, and was always ready for any mission of adventure or danger. Fred's easy good nature, and ready sympathies, gave him the key to the good will and affection of the Battery boys. He certainly ranked the associate officers in popularity.

In February, 1864, by virtue of a general order of the War Department, applying to all batteries reduced in numbers, Lieut. Dolbeer and he were mustered out as supernumerary officers. It was his intention to recruit the company to the maximum standard, and thereby to secure re-appointment. This design was defeated by the subsequent capture of the organization. "Lieut. Fred" then entered upon mercantile pursuits in his native village, Mt. Morris, Livingston County, N. Y. In this vocation he has had remarkable success.

He has committed matrimony.

Those who recollect his rabid political tenets, will not be surprised to learn that he is largely responsible for a course of lectures lately delivered in Mt. Morris, by Fred'k Douglas, Theodore Tilton, Miss Anna Dickinson, and others, cast in the angular mould of the Radical Republican. Despite of his extreme political connections, all who know Fred will freely applaud and honor his sincerity. In the hope that abundant prosperity may attend him, we leave him to make history.

86. HATHAWAY, CHAS., Perry.—Enlisted August 29th, 1862. Mustered in at Buffalo, August 30th, 1862. Was taken prisoner at Plymouth, and died at Andersonville Hospital, of chronic diarrhœa, June 12th. The number of his grave is 1,891.

Charley was one of those who, at the time of the last accession to the Battery, had made up their minds to enter the army; and whether they went with the Battery or with some other organization, was of secondary consideration.

He made a good soldier. He had entered the ranks with the right spirit. He was a pacifier at our outbursts of impatience and resentment at supposed wrongs.

He was fond of the good things of this world, and none appreciated the gifts from home more than he; but he was always ready to share with his fellows.

He saw but little of the sufferings at Andersonville, as he was the seventh man that died there. He had suffered from miasmatic fever at Plymouth, and was in no condition to undergo such a change of climate, and want of proper food.

He was serene, willing and prepared to die; and gave us the few loving words to his parents and his friends, with the calmness of one who felt assured of a heavenly home.

87. HINTON, WM. H., Rochester, N. Y.—Enlisted and mustered in October 1st, 1864, for one year. Joined November 9th, 1864, at Roanoke. Transferred to Third New York Artillery, May 25th, 1864. Was a re-enlisted veteran, having served in the two year enlistment of 1861, in infantry. Good, trusty soldier.

88. HOLMAN, GEORGE, Hague, N. Y.—Enlisted September 28th, 1861. Re-enlisted at Plymouth, January 1st, 1864. Was taken prisoner at Plymouth. We knew little of him during the prison experience. He lived through it, however, and we find him reported absent at College Green Barracks, Annapolis. Rejoined, an exchanged prisoner, in May, 1865, looking fat and healthy, and cleaner than before his capture, from which we may

infer that prison life improved him. He was mustered out at Syracuse, July 7th, 1864. Was married after his return from the army, and now lives in Albany.

89. HOLLISTER, BENJAMIN H.—After writing us pleasant words of encouragement, he says: I enlisted the 28th of August, 1862. Mustered in at Buffalo, September 10th, 1862. Was in battle at Plymouth, N. C., April 18th, 19th and 20th, 1864. In Andersonville Prison from May 1st until September 11th, when I was removed to Charleston, S. C., taken to hospital and kept there until November 28th, when I was paroled with the first ten thousand sick, and placed on board the U. S. Transport at Savannah, Ga., November 30th, 1864. Arrived in Annapolis, December 4th. Received furlough about the 20th, for thirty days, which was extended to thirty more. I then reported for duty at the U. S. G. Hospital, and in consequence of going on duty before being perfectly recovered, I was seized with typhoid fever—was unconscious thirty-one days—was then sent to U. S. G. Hospital, at Baltimore, Md., and from here I was discharged, May 23d, 1865.

Present residence, Wellsville, Alleghany Co., N. Y.—Occupation, window-blind manufacturer. Married here to Miss Alice D. Macken, and have as fine a young soldier as the country can boast of.

90. HOMAN, CHARLES H., Perry.—Enlisted, August. Mustered in at Buffalo, August 30th, 1862.

A chronic ailment caused his name to be placed on the invalid list, and he was in and out of hospitals until we reached Plymouth. He writes: "I was sent to New Berne, to the general hospital, to be treated for the chronic diarrhœa, and after I got a little better, I was sent to the head surgeon, who examined me, and his judgment was that I might do for the invalid corps, and so I was sent back to my quarters, and in a short time I and two or three hundred more, were sent north, to Newark, New Jersey, to be treated for whatever was the matter with us. I got a little better here, so that I could get out, and then I was sent up to the head-quarters again and examined, and the doctor said I would do for the second corps of invalids, and so in I went. I stayed in this some time, until I got better, and then I was examined again, and transferred to the first corps, and after staying here some time, I was sent down to Point Lookout, Md., to guard rebel prisoners. The prisoners sent here to Point Lookout, fared as well as our own boys did. I stayed here until Grant compelled Lee to surrender, and in a short time I was sent to Albany, N. Y., and mustered out of the service, after the war closed."

He subsequently adds: "I was transferred to the second battalion of the veteran reserve corps, on the 7th July, 1864, and then sent to Newark, N. J., and sometime in the winter of 1865, (January or February) I was transferred to the first battalion, Eleventh Regt., Co. G., Captain Benson Brown, and in March was sent to Point Lookout, stayed there until sent to Albany, and discharged, June 29th, and received my pay on the 7th July, 1865, and went home. Making two years, ten months and some days, that I was in the army."

Present address—Perry, N. Y.

91. HORTON, CHARLES, Albany, N. Y.—Enlisted and mustered in for one year, September 15th, 1864.
Transferred to Third New York Artillery, May 25th, 1865.
Joined at Roanoke, October 12th, 1864.

92. HOSFORD, WM. F., Perry.—Enlisted August 29th, 1862. Mustered in at Buffalo, August 30th, 1863.
At the time of his enlistment he was a student at Perry Academy. His peculiar tenacity for correctness, his taste for thorough investigation and research, his flat refusals to believe anything that could not be historically or logically proven, gave him promise of a rank among the best of scholars. He carried his love of study with him into the army, and paid but little attention to anything but his books and his duties. He was a consoler in time of trouble, and a popular prophet, inasmuch as wisdom of mind is supposed to accompany wisdom of speech.

He was taken prisoner at Plymouth. We believe that just before his capture, he was promoted corporal.

During his prison life, he was buoyant and hopeful. On account of such a spirit, we thought, and yet think, that he would have endured to the end the treatment that was there received. But he was seized with malignant typhoid fever, and although we had him removed to the hospital, and did all in our power to save him, he died, June 24th, 1864, a few days after the disease made its appearance.

The number of his grave is 2,445.

93. HOYT, WILBUR M., Brooks Grove.—Enlisted August 29th, 1862.

Mustered in, August 30th, 1862.

No nobler soldier served his country for his country's good. A man of fine physical proportions, of sound morals and integrity, and of strong religious tendencies. He had prepared himself to fill the role allotted to him: to fight and to die, with courage and with calmness. During the early part of the attack on Plymouth, his piece was ordered into action, at the first embrazure, at the right of the large house, used as the Battery barracks. Hoyt was No. 1 (that is the man using the rammer and swab). The piece had been firing some little time, when they commenced to fire more rapidly. As he was "ramming a shell home," No. 5, (whose duty it was to keep the vent hole covered) became excited, and in turning round to give some instruction concerning the ammunition, he uncovered the vent. A premature discharge immediately followed, and rammer, shell and all, went through the arms of Hoyt. One arm was shot off, the other shattered, and his face and body blackened skin deep with the burnt powder. Captain Cady standing near, was so deafened, that for a day or two, it was almost impossible for him to hear anything. One or two others were slightly injured. Hoyt fell back, but was not satisfied to be taken away until he saw his post filled and his piece at work.

He was removed to the hospital, and there, where many were shrieking with slight wounds, he endeavored to quiet them, and then urge them to the consideration of death, in a Christian's view, with resignation, hope and faith. Not a groan, not a murmur escaped his lips. He had loved his country, he had served his country, and he was willing to die for his country.

The physicians were very kind to him. The nurses looked upon his courage with wonder. Everything was done for him that could be done. He lingered along, until after we had surrendered. A few wounded were left with him, while we were marched off. He died, April 26th, 1864.

Was buried on a vacant lot west of the building occupied as the hospital of the Eighty-fifth Regiment, on the brow of the hill, overlooking the river. While in the quartermaster's department, at Roanoke, Lieut. Camp had head boards made, and after the recapture of Plymouth, by our forces, Stoddard had them erected over his grave.

94. HUBBARD, H.—Joined for duty, Nov. 16th, 1861, at Ticonderoga.

95. HUGHSON, WALLACE E., Rochester, N. Y.—Enlisted and mustered in, for one year, Oct. 1st, 1864, and joined November 9th, 1864, at Roanoke. Transferred to Third N. Y. Artillery, May 25th, 1865. From Hamlin, N. Y., was a relative of Rufus Ainsworth, and a re-enlisted veteran, having served two years on enlistment in 1861.

96. HUMPHREY, ARTHUR, Poughkeepsie, N. Y.—Enlisted, for one year, September 22d, 1864. Joined at Roanoke, October 12th, 1864. Transferred to Third N. Y. Artillery, May 25th, 1865.

97. HUMPHREY, CHAS., Kingston, N. Y.—Enlisted and

mustered in, for one year, September 29th, 1864. Joined October 12th, 1864, at Roanoke. Transferred to Third N. Y. Artillery, May 25th, 1865.

98. HUNTER, EDWIN H. JR., New York.—Enlisted February 18th, 1862. He was promoted corporal by Captain Lee, and on the 14th April, 1862, was promoted sergeant. He was very retired, but much liked by all. While we were stationed at New Berne, he was greatly shocked by reading in the New York paper of the murder of his mother, by the hand of his father. During a storm of anger, the father had buried a hatchet in the mother's skull. He was furloughed, and came to New York during the trial. He returned to the Battery, but, of course, quite changed. He often expressed the depression and the degradation that he felt. He was generally and deeply sympathized with, and, to our knowledge, he was very grateful for the many kind words and kind acts tendered him. At Plymouth he was ailing from the effects of " chills and fever," and suffered quite severely from that complaint, during the last two days of the battle. He was taken prisoner and died at Andersonville Hospital, of typhoid fever, July 15th, 1864.

The number of his grave is 3,365.

99. HURLBURT, E. T. M.—He writes :—" I enlisted in Perry, August 29th, 1862. Mustered in at Buffalo, August 30th, 1862. Mustered out in New Berne, May 23d, 1865. I was in the battle of New Berne, on the anniversary of the capture of that city by Burnside, and

in the terrible battle of Foster's Mills, and did as much retreating as any of them. I was not taken prisoner. I did not re-enlist. I was detailed as clerk in the engineer's office at New Berne, nine days before the battle of Plymouth, when our boys were taken prisoners, which detail saved me, probably, from a prisoner's death. I was promoted to corporal in October, 1862, at Newport Barracks. I was again detailed as apothecary, in Beaufort General Hospital, in July, 1864. While on this duty, was placed in medical charge of Refugee Camp, at Beaufort, and after acting in the capacity of apothecary and assistant-surgeon, for five months, was ordered to Mansfield General Hospital, at Morehead City, as a patient, but was soon ordered on duty as acting-assistant surgeon, by Surgeon Palmer, then in charge of hospital, in which capacity I continued to act until I received my discharge; the most of the time having charge of the bedded patients, though a private, for which service I received a document stating the above facts, signed by Surgeon J. C. Salter (in charge), endorsed by Surgeon D. W. Hand, in charge of Medical Department of N. C. At one time I had charge of the sick ward, while four, wearing shoulder-straps, were doctoring convalescents. This document also states that I, at one time, had charge of the General Hospital, but it was by default of Dr. Mudie, then by authority in charge.

"After receiving my discharge, I graduated at Buffalo Medical College, practiced medicine one year, at Ridgeway, N. Y., came to Hannibal, opened an office, July 12th, 1867, and have every reason to be satisfied with my choice of vocation and location,"

100. JACKSON, DANIEL, Leroy, N. Y.—Enlisted and mustered in, September 1st, 1864, for three years.

En route to join the Battery, he jumped overboard the transport, at Fortress Monroe, and deserted. For further particulars, see *Henry Raymond*.

101. JOHNSON, GEORGE B., Perry.—Enlisted October 1, 1861. Was promoted corporal. Was taken prisoner at Plymouth.

Johnson was a man quite advanced in age. He was intellectually and argumentatively strong. He was notably radical in his feelings and in his speech. While we were in garrison he perused the newspapers with eagerness and thoroughness, and few in the Battery were possessed of as much knowledge concerning the war as he. Politics were his favorite theme; though, during his imprisonment, his mind was called more to the Bible; and his conversations with his comrades indicated that that Book had received a great deal of study and attention from him.

He was sent from the stockade at Andersonville to the hospital, to do police duty. Gradually, however, he succumbed to that dread malady, chronic diarrhœa. He saw that he was declining—made his will, and a disposition of the few things he possessed—and prepared himself for the great change.

He died Sept. 21st, 1864. The number of his grave is 9,495.

102. KEENEY, GEORGE W., Perry.—Enlisted August 29th, 1862. Mustered in, August 30th, 1862, at Buffalo.

Was taken prisoner at Plymouth, and died at Andersonville, May 20th, 1864. The number of his grave is 1,250.

He was the second member of the Battery who was sacrificed at that unholy altar. His life had been unusually free from immoderate and exceptional habits; his mind was pure, and his heart kind.

The circumstances of his death were peculiar. He was taken ill with diarrhœa soon after we had entered the stockade. The surgeon prescribed an opiate, with directions, if he was no better after a reasonable time, to administer a second dose.

The man acting as nurse failed either to understand or to comprehend the surgeon's orders, and while George was still quiet, and under the effects of the opiate, the man gave him an additional quantity. George never woke from his first sleep. His sister, Kate R. Keeney, has dedicated the following lines to his memory:

"In Memoriam."

Darling, laid low in that sunny land,
  In the sleep which knows no waking,
From thy soul's high home, canst thou understand
  How our hearts are almost breaking?

A year this morn, ere the sun's first beams
  Bathed in light that Southern prison,
Thy spirit, borne from its fever'd dreams,
  To a holier life had risen.

Thy place of rest we may not see—
  Oh, God! thy aid not scorning;
We leave our darling one with Thee,
  'Till the Resurrection morning.

Peace's dewy wings again are cast
    O'er our tried, our suffering nation;
And those must be forgotten last,
    Who died for her salvation.

Spring came again, and the soft winds sighed
    To the buds, and the springing clover;
When an angel came from "the other side,"
    And beckoned our mother " over."

Ye are gathered home, our loved and lost;—
    And I stand without a shiver,
And think, as I look where the dear ones crossed,
    How little way 'tis o'er the " River."

The birds sing sweet in the homeside trees,
    And the flowers smile up to their "keepers,"
But our hearts ache on in hours like these,
    As we think of our household "sleepers."

Though my bark sails on to "the unknown sea,"
    With dirge-like gales to waft her,
Mother and Brother are waiting for me,
    In the land of the " hereafter."

HILLSIDE HOME, May 20th, 1865.

103. KEITH, G. H., Albany.—Enlisted November 16th, 1861. Promoted corporal. He died at Newport Barracks, November 2d, 1862. We find the following account in the correspondence of the " Wyoming *Times*," dated November 21st, 1862.

" Yesterday (Sunday) morning, a gloom pervaded our camp, occasioned by the death of one of the members of our company.

" His name was G. Harrison Keith. He belonged to the older portion of the company, having joined with them at Albany. His residence was in Johnstown, Fulton County. He was young and active, held the position of corporal, and was deservedly popular. He died rather suddenly. He had been severely sick with billious typhoid fever, but the day previous to his death was said to be improving rapidly. Funeral service was held in the afternoon at the chapel, and his body, followed by the whole company, was carried to Newport and buried with all becoming ceremony. This is the first death that has occurred since our arrival. The slow moving ambulance, the solemn procession winding its way through the narrow, woody road, the pines murmuring a funeral dirge as we passed under their branches, all combined, must have made even the most reckless reflect upon the brevity of life, the certainty of death, and the necessity of a preparation to meet it."

104. KELLOGG, G W., New York.—Enlisted and mustered in, November 13th, 1861. Promoted corporal by Capt. Lee. Acted as quartermaster for a short time at New Berne. Re-enlisted as a veteran at Plymouth, January 2d, 1864. Mustered out of the Battery at Plymouth, on account of promotion, in April, 1864. Promoted to second lieutenant, Third New York Independent Battery, January 8th, 1863. Promoted to first lieutenant, October 31st, 1864.

Was in all the battles, under Grant, in Virginia, including the Battle of the Wilderness, Spottsylvania, Cold Harbor and the taking of Richmond.

Is married, and now lives in New Jersey.

105. KETCHUM, RICHMOND A.—Enlisted at Rochester, N. Y., September 30th, 1864, for three years, and joined at Roanoke, November 9th, 1864.

Transferred to Third N. Y. Artillery, May 25th, 1865.

106. KING, SYLVANUS.—Joined for duty from Hamlin, N. Y., October 15th, 1861. Re-enlisted as a veteran at Plymouth, in January, 1864. Was taken prisoner at

Plymouth, and died at Andersonville, September 14th, 1864, of scorbutis. The number of his grave is 8,738.

107. KNOWLDEN, HENRY C., New York.—Enlisted and mustered in, April 12th, 1864, for three years. Transferred to Third N. Y. Artillery, May 25th, 1865.

108. LAPHAM, LEDRA H., Moscow, N. Y.—Enlisted August 29th, 1862. Mustered in at Buffalo, August 30th, 1862. Was taken prisoner at Plymouth.

It was a great sacrifice for Newton and the Lapham brothers to give up a lucrative business, pleasant homes and flattering prospects to enter the army; and yet, full of the fire of loyalty, they did it.

Ledra Lapham was not a strong man. A portion of the time, while in garrison, he seemed greatly improved, and, even while at Andersonville, he evinced that wonderful tenacity for life which we are always astonished to see in those whom we have been accustomed to look upon as feeble and delicate. He endured prison experience with but little complaint or fretfulness, while many stronger ones were uselessly querulous and disagreeable.

Newton and he were together in the hospital. It is quite noticeable that they should have been so much together, and finally died within a few days of each other.

Lapham died August 5th, 1864, of chronic diarrhœa. The number of his grave is 4,871.

109. LAPHAM, HORACE, Moscow, N. Y.—Enlisted August 29th, 1862. Mustered in at Buffalo, August 30th, 1862. Was a brother of Ledra Lapham.

Against his wishes, he was discharged from the service for physical inability. He now resides in Genesee, Livingston County, N. Y.

110. LAWLER, E., Hamlin, N. Y.—Joined for duty, October 21st, 1861.
Nothing more is known about him.

111. LEE, ABRAM, Perry.—Joined for duty, October 3d, 1861. He re-enlisted as a veteran, January 1st, 1864, at Plymouth. Was taken prisoner and sent to Andersonville.
He died at Andersonville Hospital of chronic diarrhœa, June 14th, 1864.
The number of his grave is 1,944.

112. LENT, ABRAM, Perry.—Enlisted August 29th, 1862. Mustered in at Buffalo, August 30th, 1862. Was taken prisoner at Plymouth, and died at Andersonville Hospital of pneumonia, June 29th, 1864.
The number of his grave is 2,686.
Abe Lent was long and familiarly known to the citizens of Perry. Most of us who enlisted at that time knew him well, and we recollect that many suspected that he was up to some of his shrewd tricks, and did not intend really to go with us. When we were placed in rank, in the room of the mustering officer at Buffalo, Abe refused for some time to raise his hand and take the oath. He, however, yielded to the persuasions of his friends, and was sworn. Many attributed his conduct to an unwillingness to go, but we are inclined to think that he stated his true reason, which was, that in his opinion

Hastings should have been sworn in with the rest, to carry out the understanding among the boys at the time of the enlistment. He did not understand that Hastings was to be mustered in as a commissioned officer at Albany. Familiar with bookkeeping and reports, he made himself quite useful at the battery headquarters at that kind of labor. During the latter part of our stay in Plymouth, however, he took a post at the piece, and worked well. Abe was very fond of spending an evening reviewing and rehearsing the acts and sports of the men of Perry who used to be his chums. He was happiest when he could tell us his stories of the political and other intrigues and maneuvers that he had been cognizant of, while others were in ignorance and blindness, and he chuckling to himself as he heard their innocent remarks and saw their unsuccessful tactics. Even in prison life, a retrospect of the past was his habitual resort for passing the time pleasantly. When he was sent to the hospital, he was very low, but he recognized his friends. He failed rapidly, and became delirious and insensible.

He died in unconsciousness.

113. LEONARD, FRANCIS, Albany.—Enlisted November 16th, 1861. Was appointed corporal at the time of the organization of Battery "B" of the Rocket Battalion. Re-enlisted January 1st, 1864, and was taken prisoner at Plymouth. Was exchanged and joined for duty again April 27th, 1865. He joined the Company at Coanjock Bridge early in May, 1865, looking fat and healthy, a

neat, tidy soldier. He was married after he left the service, and now lives at Albany, New York.

114. LLOYD, H. P.—He writes: "I enlisted at Angelica, New York, August 25th, 1862. Was mustered in at Buffalo, September 5th, 1862. Was mustered out of the Battery, March 11th, 1864, at New Berne, N. C., to accept promotion in the Twenty-second New York Cavalry. Was not captured, but was in hospital at Annapolis, Md., when some of the men were returned from Andersonville and Florence. Saw Sam and William Nichols—both in a very reduced state—unable to sit up. Conversed with them, and did what I could to relieve them.

"John Russell and Carnahan were there also. Some others were there, but I was not able to see them personally. I was promoted to sergeant, October 10th, 1862, and was promoted to first sergeant in December, 1862, or January, 1863. On the 12th of March, 1864, I was commissioned first lieutenant of the Twenty-second New York Cavalry, by Governor Seymour. On the 13th of July, 1864, I was commissioned captain by the same.

" On the 24th of January, 1865, I was commissioned as major in the same regiment, by Governor Fenton.

" On the 13th of March, I was brevetted by the President for 'gallant and meritorious conduct' and was recommended for brevets three other times by the Corps Commanders.

" I was detailed as a member of a Military Court of Inquiry, and as a member of two different Courts Martial, at Winchester, Va. in the winter of 1864 and 1865. In February, 1865, I was appointed aid-de-camp on the staff of Major-General William Wells, and served in his staff until active hostilities ceased. In April, 1865, I was appointed by the Secretary of War, as Commissary of Musters for the Cavalry Corps of the army of the Shenandoah, and I served in this capacity on the staff of Gen. Lorbert and Gen. Reno, until Aug. 1st, 1865, when I rejoined my regiment and was mustered out of service at Rochester, N. Y., Aug. 8th, 1865.

"I was engaged in all the battles of the Army of the Potomac, under Grant, until Aug. 1864, when our division of the Cavalry Corps was sent to the Shenandoah Valley, under Sheridan.

"On the 21st August, 1864, at Smithfield, Va., I received a severe gun-shot wound through the body and right lung, and narrowly escaped capture."

Lloyd is one of the active members and corresponding secretary of the Young Men's Christian Association of Cincinnati.

He has built up a lucrative and successful law practice.

Married June 16th, 1869, to Miss Hattie G. Raymond, daughter of John H. Raymond, L.L.D., President of Vassar College.

Present address, Cincinnati, O.

115. LOOMIS, HIRAM, Mt. Morris, N. Y.—Enlisted at Perry on the 29th of August, 1862, was mustered in at Buffalo on the 30th. Taken prisoner at Plymouth, April 20th, 1864.

Was at the Andersonville, Florence and Charleston prisons. He says: "After I had been in Andersonville about five months it was thought that Sherman was about to invade the portion of the State in which the prison was located, so they removed us to Charleston and from thence to Florence.

"When I went to Andersonville, I was sick and could scarcely keep my place in the ranks, but with Wirz at my back with a revolver pointed at me, I felt called upon to put forth every possible effort.

"Was at Florence three months.

"Was exchanged at Charleston, Dec. 10th, 1864. Was afterwards detailed as orderly in the Navy Yard. Mustered out June 28th, 1865.

"Was married to Miss Annie W. Sweetman of Mt. Morris, N. Y., October 25th, 1865."

Has been, since he left the army, quite successful in the cabinet business, at Pioneer, Williams Co., Ohio.

116. McCLAIR, JERRY.—Mustered in at Buffalo, September 30th, 1862. Was promoted to corporal. Interested himself in the recruiting of negroes, and we believe received a lieutenant's commission in a colored regiment. We never received a reply to our communication to him. Have heard that after the close of the war he was interested with Lieutenant Camp, in business in North Carolina. Gave that up, returned to Moscow, and was married. Settled for a short time in Moscow, and has now returned to the south again.

117. McCRARY, ORRIN S., Mount Morris.—Enlisted September 9th, 1862.

Was taken prisoner at Plymouth, and, as we are informed by Ferguson, died at Florence, S. C., in the Fall of 1864.

Of the three McCrary brothers who started out with us in the Fall of 1862, Charles only remains.

Orrin was a sprightly, affable fellow, ready to do and to say anything to please. As a prisoner, he had little to complain of, and cheerfully looked forward to deliverance. But the lingering, sickening delay overpowered him, and he too fell, with the thousands of others that could no longer endure their prison tortures.

118. McCRARY, WM. A., Mount Morris.—Enlisted August 29th, 1862. Mustered in at Buffalo, August 30th, 1862. Promoted corporal, November 4th, 1862.

While at New Berne, he was attacked with chronic diarrhœa, and never fully recovered. He died of that disease, August 14th, 1863.

His body was embalmed and brought North for interment. He was in so little active duty with us, that he had no opportunity for displaying his qualities as a soldier. His death was regretted and felt by all the members of the Battery.

119. McCRARY, CHARLES, Mount Morris—the third of the three brothers.—Enlisted August 28th, 1862, and was mustered in at Buffalo, August 30th, 1862. Was discharged on account of physical inability. Present address, Wellsville, N. Y.

120. McCRINK, JOHN, Perry.—Enlisted August 28th, 1862. Mustered in at Buffalo, August 30th, 1862. Was taken prisoner at Plymouth, and died at Andersonville Hospital, of chronic diarrhœa, August 19th, 1864. The number of his grave is 6,203.

John's tongue wagged ceaselessly. His body might be wearied, his spirits subdued, but his tongue never cared for rest. He was pretty well advanced in years; had had a good deal of experience as a traveler, and his stock of stories was large; and if they needed a little burnishing, he had quite a faculty for inventing additional occurrences which should keep up the interest of his hearers.

In prison he kept up good spirits; and if a body was not too dejected, an hour's interview with him was a relief from the more sedate and quiet comrades.

After he reached the hospital, he became greatly alarmed, as he realized that death was approaching. Several nights in succession we were roused at midnight by a message from him that he was dying, yet he lingered along for some time. In the daytime he would seem to be improving, and at night would fail. We managed at last to obtain a Roman Catholic priest, who made him a visit, and the comforting assurances which John received from him seemed to quiet his alarm, and he died, being himself hardly aware that he was breathing his last.

121. McCrink, James, Perry.—Enlisted December 22d, 1863. He reached Plymouth just in time to be taken prisoner.

We are unable to trace him any further, but it is supposed that he died in prison.

122. McDonald, Archibald, New York.—Enlisted November 26th, 1861. Re-enlisted as a veteran, January 1st, 1864. At one time acted as orderly sergeant. Was promoted corporal in 1864.

Was taken prisoner at Plymouth, and died at Andersonville Hospital, of typhoid fever, September 15th, 1864. The number of his grave is 8,969.

He was a faithful soldier and a willing worker, a radical thinker and a plain speaker, yet a practical promoter of obeyance to orders and strict discipline. To the oldest portion of the Battery he was best known, and was quite popular with them.

123. McEwen, George W., Ticonderoga.—Joined for duty October 2d, 1861. He was in poor health most of the time, and we conclude, as his name cannot be found on the later muster rolls, that he was discharged from some of the hospitals. Re-enlisted, as a veteran volunteer, at Plymouth, January 1st, 1864,. Received veteran furlough and never returned to his company. Had charge of the cook house when we first enlisted the two colored cooks, allowed by law (George and Nelson).

Little *Pete* coming up to the quarters, one day, sung out to McEwen—" Ho ! Mack, the boys go back on your nigger cooks, ha, ha ; dats too bad, ha, ha." Mack replied, " O, you d—d black imp, what you talking about"—at the same time picking up a stone to throw. Pete replied, " Oh, Mack, oh, Mack, I didn't mean nothing, the're all right. I'd just as leave eat after their cooking as after yours."

124. McGuire, Thomas, Gainsville, N. Y.—Joined for duty, October 1st, 1861. Re-enlisted, as a veteran, in January, 1864.

Was taken sick while on his veteran furlough, and arrived at Roanoke Island, April 20th, 1864, just after Plymouth was captured. He remained with the remnant of the Battery until it was mustered out in Syracuse.

Present address, Gainsville, N. Y.

125. McGuire, James, Gainsville, N. Y.—A brother of Thomas; joined for duty, February 24th, 1864. Taken prisoner at Plymouth, April 20th, 1864.

126. McGuire, Michael.—Enlisted March 21st, 1864, at Gainsville, N. Y., for three years; joined April 20th, 1864; died at hospital, on Roanoke Island, August, 1864, of acute dysentery. Is buried near the hospital (brother of Thomas McGuire).

127. McNinch, Henry, Moscow.—Enlisted December 19th, 1863.

Was taken prisoner at Plymouth, and is reported by Newcomb to have died at Florence.

He was one of the last recruits before the battle of Plymouth, and we do not know much about his movements after he reached Andersonville, as we had little acquaintance with him.

128. McVey, James.—Mustered in about Sept. 30th, 1862.

McVey was one of the young men that volunteered from Hamilton College, at the time George Hastings joined the Battery. He was a restless fellow, and felt the restraint of army discipline severely. On this account he made a poor soldier. He was talented, had a keen sense of honor, and to our thinking, too high an appreciation of caste. After a little time he was put on detailed service, in the General Department, at New Berne. Was promoted to a lieutenant's commission in the Third New York Artillery. Was aid-de-camp on General Peck's Staff. Came with General Peck to

New York City, and remained on his staff while he had command of that department. He was there mustered out of service. We have been told that he returned to his home, in the interior of New York State, and died there.

129. MAREAN, CHARLES A., Moscow.—Enlisted August 28th, 1862.

Mustered in at Buffalo, April 30th, 1862.

He was quick and impulsive in his enlistment.

As he was quite young and inexperienced, he could hardly realize his undertaking; yet, in actual experience he proved himself steady and capable. He was taken prisoner at Plymouth, and we are informed by Ferguson, died at Florence, S. C.

130. MARRIN, PATRICK, Perry.—Enlisted November 21st, 1861. Re-enlisted as a veteran, January 1st, 1864. Was taken prisoner at Plymouth. Marrin was very severely wounded at the battle of Plymouth, being struck with five Minie bullets, while at his post in charge of the caisson teams. One passed through his hat, just grazing the skin; two bullets, not five minutes apart, passed through the fleshy part of his legs, above the knee; another lodged in his ankle, and remains there yet, causing him a great deal of suffering at times. He showed no cowardice, nor flinching, but remained at his post as long as he was able. After he was wounded, he started for the hospital, using two pieces of palings as crutches, and on his way was met by some rebels who stopped him, set him down, pulled off his boots, took his hat, and then set him up and let him go on minus hat and boots, which were too good to lose.

He remained at Plymouth some time, with others of the wounded, and was finally sent to Andersonville.

His wounded leg troubled him considerably, and he was sent to the hospital. He there made himself so useful that he remained as an attendant until he was exchanged.

He was paroled in November, 1864, at Savannah. Joined after exchange, at Coanjock, in May, and was transferred to Third New York Artillery.

Present address, Perry, N. Y.

131. MARRIN, CONNOR, Perry.—Enlisted November 21st, 1861.

Was discharged from the hospital on account of physical inability. The following has been copied from a newspaper (name and date not stated) " Connor Marrin, a resident of Perry, and a member of Lee's battery, at New Berne, N. C., returned home on Monday evening. He has dropsy on the liver, and has received his discharge in consequence."

We have heard that he was with his brother in California.

132. MARTIN, HECTOR C., Warsaw.—Enlisted, October 12th, 1861.

Was mustered in as bugler, and held that position for some time. It did not suit his taste, however, and others, whose musical genius better fitted them for the position, were appointed in his place. He was promoted quartermaster sergeant, November 4th, 1862.

There is no evidence in the muster rolls that he re-enlisted as a veteran.

If our memory serves us rightly, he had determined to serve out his three years, and then return to his family. He was taken prisoner at Plymouth, N. C., and died at Andersonville Hospital, August 7th, 1864.

The number of his grave is 5,086.

133. MEADE, GEORGE F. II., Moscow.—He was mustered in some time in 1864.

We cannot find his name in any of the muster rolls in our possession.

William Carnahan writes that Meade was shot in the battle at Plymouth, and that he saw him after he was dead. We believe that he was attached to Williams detachment, and the current report in the story of the battle of Plymouth, as told by the men of that detachment, was, that Meade was shot dead instantly. The ball passing either through his head or his heart.

134. MERRILL, J. W., Perry.—Enlisted August 30th, 1862.

Mustered in at Buffalo, on the same day.

Was appointed sergeant, November 4th, 1862.

Was reduced February 18th, 1864, to private, by his own request, in order to enable him to accept a detail in the quartermaster's department, at Plymouth. At the battle of Plymouth, by request of Captain Cady, he resumed the command of his old detachment, which was stationed at the extreme right of the line of works. While in the army was correspondent for "The Wyoming Times" and "The Western New Yorker," over the signature of "J. W. M."

Was discharged from the service, by special order of the Secretary of War, No. 157, on the 20th of April, 1864, in order to allow him to accept a commission as second lieutenant in the Second New York Artillery. (See vol. 2, New York State Adjutant's Report, 1868.) At the same date, and before the discharge and commission had reached him, he was taken prisoner at Plymouth. During his imprisonment he remained a month in the stockade. Was sent from there to the hospital. By the kind attention received from Dr. A. W. Barrows, of Amherst, Mass., a fellow prisoner, he sufficiently recovered to do duty in the hospital, in caring for the sick. Was a short time in Millen prison. Was paroled for exchange at Savannah, November 20th, 1864. He reached the Federal steamer with little clothing, penniless and hungry. Remained in General Mulford's office, on the flag ship, "New York," of the exchange fleet, a month. Was mustered out of service, August, 1865, in New York City.

Was in the Treasurer's office of the United States Telegraph Company, about a year. In March, 1866, was elected Secretary of the "North America" Life Insurance Company. On the 1st of May, 1869, was elected Vice-President of the same company, vice T. T. Merwin, resigned, and was sent to California, partly to visit the "Pacific Branch Agency" of the "North America," and partly for his health.

Married Miss M. C. Morgan, of Brooklyn, April 25th, 1867.

Present address, 17 & 19 Warren Street, New York City.

135. MILLER, GEORGE, Hamlin, N. Y.—Joined for duty October 23d, 1861.

Re-enlisted at Plymouth, January 1st, 1864. Reported on company roll as "absent at College Green Barracks, Annapolis, as a paroled prisoner."

136. MINER, J. GILE, Perry.—Enlisted October 5th, 1861. He re-enlisted as a veteran, January 1st, 1864. Was taken prisoner at Plymouth, and died at Andersonville Stockade, of chronic diarrhœa, August 3d, 1864. The number of his grave is 4,771.

Gile was the sutler of our camp, he was a hard worker and did not allow his store to interfere with his duties. He was shrewd—understood when and where to accommodate with finances—and withal, managed to keep his matters of business so to himself, that few knew how much or how little he accumulated.

We did not see him after we arrived at Andersonville, and cannot therefore say anything of his prison experience.

137. MOSIER, MARION R.—Enlisted in Weathersfield, Wyoming Co., N. Y. Mustered in at Buffalo, October 22d, 1861. Mustered out at Elmira, N. Y. Re-enlisted as a veteran at Plymouth, January 1st, 1864. Was prisoner at Andersonville, Charleston and Florence. Was paroled at Wilmington, N. C. Married, May 11th, 1865, to Rosettie Lewis, of East Pike, N. Y.

Present residence, East Pike, Wyoming Co., N. Y.

138. MUNROE, DARIUS, Hager, N. Y.—Joined for duty, September 28th, 1861.

139. MURRAY, WM. R. New York City.—Enlisted November, 1861, in Company A., of the Rocket Battalion. Was transferred to Company B, and promoted orderly sergeant.

In January, 1862, becoming dissatisfied with some of the appointments, he left the Company and went to New York City, where he remained until June 2d, 1864.

He was court martialled for desertion, September 12th, 1864, and sentenced by Court Martial Order No. 50, Head-quarters, Military Governor, Alexandria:— " with loss of all pay and allowance due to him—to forfeit $10 per month for 20 months, and make good all time lost by desertion."

His sentence was upon his petition and explanation of the circumstances, revoked in part.

He rejoined for duty, November, 22d, 1864—having had a severe experience in the prisons—and was promoted sergeant, April 1st, 1865.

His promotion to the first sergeancy was heartily endorsed by all. In his latter experience with the Battery, he proved to be one of the best soldiers in the Battery.

If being deceived and misinformed by superior officers, is a sufficient cause for a soldier to desert, he had good reason for doing as he did.

140. NEWCOMB, L.—He writes: "Enlisted at Perry, October 12th, 1861, by J. E. Lee.

" Mustered in at Buffalo, October 26th, 1861.

" Re-enlisted at Plymouth, January 1st, 1864.

" I was mustered in as bugler, on account of my age, as the mustering officers would not take any person under eighteen years of age, except as a musician.

"Was appointed corporal, June 22d, 1863, and sergeant, February 18th, 1864.

"My commission as second lieutenant, Twenty-fourth Independent Battery, N. Y., dated from January 10th, 1865. The Company was transferred to the Third New York Artillery about May 27th, 1865—commanded by Col. Charles A. Stewart—and was known as Battery 'L.'

"I had my commission transferred to the Third.

"Received a commission as second lieutenant in the Third, issued the 21st June, 1865, by Gov. Fenton, to date from March 17th, 1865, but did not muster on account of an order stopping the mustering in of any more officers in that department.

"I also received a third commission as first lieutenant, issued July 5th, 1865, to date from the 2d July, 1865, but did not muster. I acted as lieutenant all the time after I joined the Company, until I was mustered out.

"I was captured April 20th, 1864, and sent direct to Andersonville; went into the stockade, May 1st 1864. I received a reprimand from *Captain Wirz* before I entered the prison; I was just recovering from a shake of the ague, and being weak, I sat down while they were taking our names at the cars,—he saw me, and said "G—d—you, get up from there, I will learn you to stand in line when I tell you, before you have been here long, you d—sons of b—;" I stood up, and think all the rest did so.

"I was put into the Thirty-ninth Detachment but soon consolidated the Detachments, and brought our number down to the Twenty-first Detachment.

"I remained in that Detachment until I left Andersonville.

"There were eight of us boys in our tent (a blanket stretched over a pole). They all went to the hospital, and three or four out of the eight died. I left Andersonville for Charleston, S. C. about September 10th, 1864, was kept at Charleston about one month. I had the scurvy, when I left Andersonville, in the right limb so badly that I could not straighten it. Went to the hospital at Charleston, stayed three days in the rain without any shelter, and finally went back to camp and thought myself better off. Left for Florence about the 10th of October, 1864, remained there until about the 18th of February, when we were sent to Wilmington, N. C., (I was just getting over the fever at this time) we could hear our forces fighting, a few miles from town. Was sent to Goldsboro', N. C. Remained there until the night of February, 25th, when we were sent back to Wilmington, to be paroled. Arrived there on the afternoon of the 26th February. I had thus been 10 months and six days in the hands of the Rebels, and left there in a few days for Annapolis, Md."

Newcomb accompanied the section that participated in the battles of Kinston, Goldsboro', &c. During the engagement at Whitehall, he dismounted and relieved one of the cannoneers, and in several ways displayed great coolness and bravery. His promotion to sergeancy followed his conduct at this battle. In speaking of the battle of Whitehall, he says :

"There were about 34 privates, cannoneers and drivers on the march. J. Button No. 2, Third Detachment, was behind from some cause, and I occupied his place. I went down to the stream for water in front of our skirmishers, and I could see the Rebel skirmishers across the stream. I was not with Bob Turner when he was killed, therefore I could not give the particulars of his death. I got the fragment of shell that he was killed by—and let Lieutenant George Hastings take it after reaching New Berne."

Mustered out of service, July 7th, 1865, at Syracuse.

Married Miss Aurelia Austin.

Present address, Perry Centre, Wyoming Co. N. Y.

141. NEWTON, RILEY J., Moscow.—Enlisted August 29th, 1862.

Mustered in at Buffalo, August 30th, 1862.

Was promoted corporal, April, 1864.

Taken prisoner at Plymouth, and died at Andersonville Hospital, of chronic diarrhœa, July 31st, 1864.

The number of his grave is 4,469.

At the time that Newton enlisted he had every right to expect a very prosperous business if he remained at home, but comprehending that the war was not to be of such short duration as many anticipated, and feeling that he was an able-bodied man, he could not persuade his conscience that it was right for him to stay away from the scene of action. He stopped building—put away the lumber—and, as in olden times, the farmers left the plough in the field, while they put the musket on their shoulder and fell into the ranks, so he dropped his tools, threw aside his work and stepped into our ranks; promising, with an understanding of what he promised—to fight with us for three years. In prison he progressed very well. In the hospital he was well cared for. He died of disease, not of starvation. In character he was positive, but mild and true. In his sickness he realized that he was low, and was prepared for the worst.

142. NICHOLS, SAMUEL.—Enlisted October 11th, 1861, at Clarkson. Re-enlisted as a veteran, January, 1864, and was taken prisoner at Plymouth. Died at the United States General Hospital, Annapolis, Md., December 21st, 1864, with chronic diarrhœa. He was one of the noble specimens of manly beauty—six feet and over in height—well proportioned, and always glorying in his strength and activity. The severity of the prison life at Andersonville, made such changes in him, that when the writer found him on the decks of the exchange steamer at Charleston, he could hardly recognize him. His death is solely attributable to that prison treatment.

143. NICHOLS, WILLIAM P.—Enlisted at Hamlin, N. Y., November 9th, 1864.

Re-enlisted as a veteran, at Plymouth, in January, 1864.

Was taken prisoner at the battle of Plymouth.

Ferguson says that "William Nichols died at Charleston, S. C."

Lloyd says that "he saw him in a very reduced condition at Annapolis Hospital."

Camp reports that "he was remarked upon in the muster roll as 'absent at the U. S. General Hospital, Annapolis, Md.,' where he was paid for September and October, 1864."

We are inclined to think that he died at Annapolis.

144. OTIS, FRANKLIN D., Hamlin.—Enlisted October 21st, 1861.

Was appointed corporal at Albany, and was one of the few who retained his position throughout all the changes of the organization. He died at Plymouth, of a congestive chill, the result of a long siege of fever and ague. He was a professor of religion, and an upright, conscientious young man. He possessed the esteem and respect of all who knew him.

145. OTIS, CHARLES, Royalton, N. Y.—Enlisted and mustered in October 12th, 1864, for one year. Joined November 9th, 1864, at Roanoke.

Transferred to Third New York Artillery, May 25th, 1865.

146. O'DELL, THOMAS.—Enlisted at Tarrytown, N. Y. for one year, October 13th, 1864. Joined for duty at Roanoke, December 1st, 1864.

Transferred to Third New York Artillery.

147. PAGE, H. C.—Enlisted at Perry, October 1st, 1861.

Mustered in at Buffalo, October 26th, 1861.

Page took an active interest in enlisting men at the time of Captain Lee's organization of the Company.

He was warranted as quartermaster's sergeant, and proved an efficient man in the right place.

Was correspondent for the " Wyoming *Times.*"

He writes:—" Discharged at New Berne, July 1st, 1862, for disability, having contracted fever.

" Again enlisted in New York City, November 23d,

1863, and mustered out with the Company at Syracuse, July 18th, 1865. Was in Andersonville and Millen prisons.

"Reached Andersonville, May 1st, 1864. After being confined in the stockade about one month, was employed by the Confederates to assist in the hospital, outside of the enclosure—to which fact I attribute the preservation of my life—was employed about the Dispensatory.

"While in the Hospital, I attended many of the Company, and saw several die.

"Paroled at Charleston S. C., November 20th, 1864, and exchanged while at Parole Camp, Annapolis.

"After being paroled, remained on the Flag of Truce boat "New York" one month, as clerk for Col. Mulford, Commissioner of Exchange. Reached Parole Camp, Annapolis, December 20th. Furloughed for 30 days, in common with returned prisoners. Returned to Parole Camp, January 23d, 1865. Forwarded to Roanoke Island, N. C., where the remnant of the Battery was stationed, under command of Lieutenant Camp, and did duty with the Battery, until mustered out as above.

"Was quartermaster sergeant from the organization at Buffalo, until discharged at New Berne. Promoted from private to quartermaster sergeant, February 1st, 1865, by Lieutenant Camp.

"Have resided in Missouri and been engaged in teaching, since leaving the army.

"Reside now at Maysville, De Kalb Co., Mo., and am practicing law."

148. PAGE, WILLIAM N.—Was detailed in quartermaster's department at New Berne.

Came North in April, 1863, and organized a Company for the Eleventh Artillery. Was promoted to second lieutenant in the Fourth Artillery. Then completed his theological studies, and during 1867, visited Europe. On his return took charge of the Presbyterian Church at Trumansville, N. Y.

In December, 1868, received a call to preach in Jacksonville, Florida, which is his present address.

Married, September 27th, 1862, to Miss Jennie A. Peck, of West Bloomfield.

149. PARMLEE, O. G., Hamlin, N. Y.—Joined for duty, November 9th, 1861.

Re-enlisted as a veteran, at Plymouth, in January, 1864.

For some reason, he did not reach Plymouth in time to participate in the battle.

He rejoined the Battery at Roanoke Island, was transferred to Company "L," Third New York Artillery, and mustered out with the rest of that Company.

150. PATTERSON, WILLIAM, West Sparta.—Joined for duty, February 19th, 1864. Mustered into service at Canandaigua, February 25th, 1864.

Taken prisoner at Plymouth. We find him reported on the muster roll as "Absent at College Green Barracks, Annapolis—a paroled prisoner."

151. PERKINS, JAMES W., Cuylerville, N. Y.—Mustered in at Buffalo, August 30th, 1862. Was taken prisoner at Plymouth, April 20th, 1864. Died at Andersonville Hospital, of chronic diarrhœa, August 28th, 1864. The number of his grave is 7,172.

In camp he went by the sobriquet of "Peter." He was full of life and fun. In his prison experience we saw but little of him, until he came to the hospital; such was his condition, at that time, that little of his former spirit was visible.

152. PHELAN, CHARLES T., New York City.—Mustered in, September, 1861.

Re-enlisted January 1st, 1864. Promoted to corporal.

Taken prisoner at Plymouth, remained in the Prison Stockade. Was exchanged December 4th, 1864, on parole.

Was sick only three days during his imprisonment.

Was at the Camp of Parole, at Annapolis, until the 9th of June, when he was discharged. Visited Johnstown, Fulton Co., N. Y., remained there until August; then went to Eastman's Commercial College, at Poughkeepsie. Graduated the 23d of December. Went into business at Poughkeepsie. On the 20th of April, went to the Island of Cuba, remained there a year and then returned to New York.

Was married to Miss Avis Dater, of Poughkeepsie, July 6th, 1868, and is now living in New York.

153. PIPER, GEORGE W., Perry.—Was mustered in at Buffalo, August 30th, 1862.

Taken prisoner at Plymouth and died at Andersonville.

We do not know the date of his death; it would not be difficult to tell the cause. We believe that he left a wife, who resides in Pike.

154. PIPER, A., Perry.—Joined for duty in February, 1864.

He reached Plymouth in time to take part in the battle, and to be taken prisoner.

He died at Andersonville about the same time that his brother died.

155. PRATT, PHILANDER, Perry.—Mustered in at Buffalo, August 31st, 1862.

Was taken prisoner at Plymouth; taken to Andersonville, and died at that place, August 21st, 1864, of chronic diarrhœa.

The number of his grave is 6,455.

Pratt was an excellent cannoneer, ready for duty and quick at his work. A quiet and pleasant comrade. He was one of the useful men at the sawmill at Newport Barracks.

We believe that in the later days at Plymouth, he was promoted corporal.

156. PRINCE, WILLIAM.—Enlisted October 4th, 1864, at Rochester, N. Y., for one year. Joined at Roanoke, November 22d, 1864.

Transferred to Third N. Y. Artillery.

157. PURDY, S. R.—Enlisted at Kingston, N. Y., September 29th, 1864. Transferred to Third New York Artillery, May 25th, 1865. Joined for duty at Roanoke, October 18th, 1864.

158. QUINN, JOHN, Perry.—Joined for duty, November 21st, 1861.

At Washington, where Battery "B," of the Rocket Battalion, was embarking on the vessels for New Berne,

Quinn defended one of the Battery boys who was light and small, in an altercation with a stronger man, a soldier of another regiment. The soldier drew a knife and stabbed Quinn several times; but, notwithstanding this, Quinn continued to fight until he had taken the knife away from his antagonist, and in turn given him several dangerous plunges of the weapon. Upon being separated, both were found to be dangerously wounded, and were removed to the hospital.

Quinn never returned to the Battery.

We have heard that he was residing in Portage.

159. RANKIN, ERASTUS.—Enlisted at Rochester, October 7th, 1864, for one year. Joined at Roanoke, December 1st, 1864. Transferred to Third New York Artillery.

160. RATHBONE, SYDNEY S., Perry.—Enlisted October 3d, 1861. Was discharged some time in 1862, for physical inability.

His historical picture, as represented by the older portion of the Battery boys, was that of a "Jolly old Ambulance driver."

161. RAWSON, PORTER D., Perry.—Enlisted August 26th, 1862.

Mustered in at Buffalo, August 30th, 1862.

Was appointed artificer, November 4th, 1862.

Brought up in a radical school, he believed in showing practically his political tendencies. He left his family and a happy home, to share the privations and the sufferings of his fellows, who were fighting out the principles which they believed were right. He was an eccentric genius, and adapted himself to his army life with little complaint. He was ready to mend or make anything named in the Artillery Vocabulary. He undertook the management of the engine in the steam saw mill, with as much assurance as if engineering was his profession; and he was one of the principal aids in making it a success. He was taken prisoner at Plymouth. From all the information in our possession, we are led to believe that he died on the cars, while being conveyed from Florence to Charleston. Ferrin saw him taken out of the Florence Hospital in a very weak condition, to be transported, with others, to Charleston, for exchange. We believe there is no further knowledge of his existence among the surviving members of the Battery.

162. RAYMOND, HENRY.—Enlisted in Second New York Volunteers, in April, 1861, and discharged with his regiment, May 26th, 1863. September 7th, 1864, re-enlisted at Albany, for one year, as a recruit for the Sixth Heavy Artillery Regiment. Was sent to Hart Island rendezvous, and there, contrary to his wishes or enlistment, was transferred to the Ninth Heavy Artillery, and in company with Daniel Jackson and others, he was forwarded to join his company on the James River, Va. After Jackson deserted, he (Raymond) determined to take his name, and answer to Jackson and come on to the Twenty-fourth Battery, instead of going to the Ninth Heavy Artillery. This he did, and was known as Jackson. Soon after Camp took command, he received a letter from the Secretary of War, enclosing a letter

REBEL MODE OF CAPTURING ESCAPED PRISONERS.

from Raymond's father to the President, stating his case and asking pardon, as he presumed he was reported as a deserter. At this time, Jackson alias Raymond, was company clerk, and anxious indeed was he to hear his fate. He had enlisted for one year, and Jackson for three years; he had stepped into the wrong man's boots, and was anxious about the two extra years. The case was kept a profound secret and not known in the Company until May, when it left Coanjock Bridge for New Berne to be transferred. Orders were received to send him to the Sixth Heavy Artillery, and with a recommendation for pardon in case of Court Martial. He was started as ordered, and afterwards it was learned that he was not court martialed but found a good company and kind officers.

Was an excellent soldier and good company clerk.

163. RICH, THURMON, Hague, N. Y.—Joined for duty, September 21st, 1861.

Re-enlisted as a veteran in January, 1864. Married while on furlough. Was taken prisoner at Plymouth, and died at Andersonville Stockade, July 8th, 1864, of chronic diarrhœa.

The number of his grave is 3,077.

164. RICHARDS, ELIAS, Perry.—Mustered in, August 30th, 1862.

Was at the second attack of New Berne, and battle of Plymouth. Promoted to corporal, by Captain Cady, at Plymouth. Was taken prisoner and was sent to Andersonville; became sick, and was sent from the stockade to the hospital, where he recovered sufficiently to aid in caring for the sick. Was paroled at Andersonville, and sent from there to Vicksburg, by way of Columbus, Ga., and Montgomery, Ala. Was exchanged at St. Louis. From St. Louis went directly to Annapolis, Md., and was finally mustered out, the 12th of July, 1865, at Rochester, N. Y.

Has since been in the employ of the Erie Railroad Company, at Hornellsville, N. Y., comfortably settled. Present address, Hornellsville, N. Y.

165. RICHARDS, ALBERT, Perry.—Enlisted October 1st, and was mustered in at Buffalo, N. Y. October 1st, 1861.

He received a warrant as artificer, in October.

In February, 1862, he accompanied Captain Lee and sister, on a visit to the Bull Run battle field, a description of which was written by a correspondent of the "Wyoming *Times*."

While on a scout out of Newport Barracks, he discovered the saw-mill which was afterwards, under Lieutenant Cady and his engineers, made so useful to the Company.

Was taken prisoner at Plymouth. Remained in the stockade at Andersonville until September 12th, 1864. Was taken from there to Charleston, thence to Florence.

While staying there, food was very scarce, and for three consecutive days, he had had nothing to eat of any kind. During his entire stay there, he had meat but three times.

Was paroled on the 8th of December, 1864. Was sent to St. John's Hospital, at Annapolis. As soon as he was in condition, he received a furlough to go home.

Remained home, quite ill, until April 7th, 1865. Then reported himself to the hospital again. Was sent from there to Camp of Parole, and finally, was ordered to his Company at New Berne. Was mustered out at Syracuse, N. Y., July 7th, 1865. Has since resided in Perry, N. Y.

166. RICHARDSON, ORLANDO, Moscow.—Enlisted August 18th, 1864.

Transferred to Third New York Artillery, May 25th, 1865.

Joined October 17th, 1864, at Roanoke Island.

He was a queer specimen of humanity. In the warmest days of July, he would wear two suits of clothes to keep warm. Sergeant Russell had him under his especial care with instructions to make a soldier of him if possible; but although Russell had had fourteen years experience in the Regular Army, he'd more than found his match in Richardson.

167. ROACH, WILLIAM, Gainsville, N. Y.—Enlisted March 24th, 1864.

Transferred to Third New York Artillery, May 25th, 1865.

Joined at Roanoke, April 20th, 1864. A good reliable soldier.

168. ROOD, LE GRAND D., Perry.—Enlisted August 28th, 1862.

Mustered in at Buffalo, August 30th, 1862.

Taken prisoner at Plymouth, and died at Andersonville, of chronic diarrhœa, June 7th, 1864.

The number of his grave is 1,735.

He was the fourth member of the Battery who died at Andersonville. While we were stationed at Plymouth, there was quite an interest aroused in the minds of several of the men, as to their spiritual condition.

Arguments upon certain portions of the Bible, had led them to a more thorough investigation of its truths.

Among those who were earnest and zealous advocates of a more faithful obedience to its laws, were Hoyt, Eastwood, Bachelder, Shirley and Rood. Rood kept up that interest, and to the day of his death, endeavored to act and speak as would become one who wished and hoped to enter the promised land of joy above.

169. ROOT, HIRAM.—Enlisted October 10th, 1864, at Rochester, for one year. Joined for duty at Roanoke, November 17th, 1864. Transferred to Third New York Artillery.

170. ROOT, STEPHEN, Hamlin, N. Y.—Enlisted October 13th, 1861.

Re-enlisted as a veteran, at Plymouth, in January, 1864.

Was taken prisoner at Plymouth, and is reported to have died at Florence, S. C.

171. ROWELL, SOLON, Clarkson, N. Y.—Enlisted, October 4th, 1861.

He was at Newport Barracks when the recruits came there in October, 1862. He soon after received a furlough on account of sickness. He never returned to the Battery.

We have been told that he was discharged at the hospital at Rochester, and is now living at Hamlin.

172. RUSSELL, ENOCH J.—Enlisted at Rochester, October 6th, 1864, for one year. Joined at Roanoke, November 9th, 1864. Transferred to Third New York Artillery.

173. RUSSELL, JOHN A., Ticonderoga.—Enlisted October 15th, 1861.
Re-enlisted as a veteran in January, 1864.
Was taken prisoner at Plymouth. He endured to the end the prison treatment, and was paroled.
Joined the Company in May, at Coanjock Station—clean, fat and healthy, never looked better.
Transferred to Third New York Artillery, May 25th, 1865.

174. RUSSELL, JOHN, Poughkeepsie, N. Y.—Enlisted September 27th, 1864.
Promoted corporal, November 18th, 1864.
Promoted sergeant, November 25th, 1864.
Transferred to Third New York Artillery, May 25th, 1865.
This man was wounded at the battle of Bull Run, 1861, where he had a part of his skull taken out, and drew a pension of $8 per month. Had served fourteen years in the regular service and felt away from home when out of the army. Thorough in discipline, he became a valuable non-com., always making the men keep themselves and their quarters in regular army style; acted as orderly part of the time.

175. SACKETT, WALTER, Albany.—Enlisted September 20th, 1864.
Transferred to Third New York Artillery, May 25th, 1865.
Joined October 12th, 1864, at Roanoke.

176. SAFFORD, PEMBROKE J., Perry.—Enlisted August 28th, 1862.
Mustered in, August 30th, 1862, at Buffalo.
Was taken prisoner at Plymouth and died at Andersonville Hospital, of chronic diarrhœa, June 12th, 1864. The number of his grave is 1,880.

It is noticeable that Rood and Safford, who represented that portion of Perry called "Buffalo Corners," in the Battery, should (after having been playmates, schoolmates, and finally comrades in battle) have died within five days of each other.

Safford never appeared like a strong man; and yet he was able to endure a good deal of hardship.

If we may judge from the warm clothing and other things of comfort, that were sent to him from his home, we should conclude that they also thought him none too sturdy. He did not stay long with us after we reached prison; and if it had been ordained by a wise God that he must be sacrificed, we think, what a kindness was there in making his stay in such a horror, brief.

177. SANFORD, L. J.

178. SECOR, ANDREW J., Rochester, N. Y.—Enlisted March 24th, 1864.

Transferred to Third New York Artillery, May 25th, 1865.

A very good soldier. Punished once by knapsack drill one hour each day for a week, under Sergeant Russell, for robbing a setting hen of her eggs, which was "against orders."

179. SHANK, LABAN H., Mount Morris.—Enlisted and mustered in, August 30th, 1862, at Buffalo, N. Y. Was taken prisoner at Plymouth, and died at Andersonville Hospital, of chronic diarrhœa, August 13th, 1864.

The number of his grave is 5,645.

Shank was a carpenter by trade, and was a man of utility in the Battery. He stood the test of a soldier's life with great endurance, until he reached Andersonville, but that proved too much, and his physical force yielded to the pressure sooner than many of the others.

180. SHELL, JOHN, Clarkson, N. Y.—Joined for duty, October 10th, 1861.

Discharged for physical inability, April, 1862.

181. SHEPPARD, NELSON.—Enlisted May 11th, 1863, at Plymouth, N. C.—a colored cook. Was taken prisoner at Plymouth, April 20th, 1864. Put in prison there and forced to trade hat, boots, watch, &c., with the rebels.

Owing to the fact that he had acted as guide on several raids our cavalry had made up towards Williamston, the inhabitants of that vicinity tried to see what they could do for Nelson, to repay his kindness. They accordingly got an order to whip him, which was done in the most approved style; the next day they dished up another dose, and for several days poor Nelson had to undergo chastisement. Was then put in a gang with ball and chain, and sent to Tarboro, N. C., where he got rid of the ball; was then sent to Weldon, to work on fortifications; there he got rid of his chain and made his escape, joining the company at Roanoke Island, in the fall of 1864.

Was transferred to Third New York Artillery.

Nelson was very shy of rebels after his treatment at Plymouth. Whenever there was talk of rebels at Coanjock, Nelson took his post near the swamp and kept his eyes pealed.

182. SHIRLEY, PHARES, Perry.—Enlisted September, 1862.

Mustered in at Buffalo, September 10th, 1862.

Was taken prisoner at Plymouth, and was sent to Andersonville. He was detailed from the stockade to do duty in the hospital. On the morning of May 21st, he came up and made quite a long visit to the writer.

Soon after his return to his own tent, he was sitting on his bunk, conversing with some of his comrades, when he suddenly fell over on his couch, and immediately expired. There was no *post mortem* examination, but he undoubtedly died of heart disease. It was a sudden shock and sad calamity to his surviving comrades, for he was uniformly kind and attentive to them all.

The following appropriate obituary appeared in the *Western New Yorker*, written by Rev. J. R. Page:

"It is not fit to suffer the worthy dead to go down in silence to the grave—to make no note of their departure—and withhold the meed of praise due to their exemplary lives and their precious memory. In this connection, I want the privilege of paying a brief tribute of friendship, in your columns, to one of our noble Perry soldiers, who died a prisoner in rebel hands.

* * * * * * * *

Born in our village (Perry) and having spent nearly all his life in it, Phares was widely known and universally regarded as one of our most promising young men. Gentle and accommodating in disposition, proverbially truthful and upright in speech and act, free from the sins to which young men in particular, are so fearfully exposed, he was a general favorite in our community, and, it is believed, he he has not left an enemy among all who knew him. Favored with a naturally amiable disposition, grace had made it increasingly attractive, and his profession of Godliness was habitually honored in practice. He was a member of the Presbyterian Church, where his funeral sermon was preached last Sabbath, by the pastor, from 1st Kings, chap. II, verse 2. The theme of the discourse was the dying soldier's legacy to each of his countrymen—' his unfinished work.' "

A copy of a letter to Phares' mother gives more particulars of his death. It is as follows:

"ANDERSONVILLE, Nov. 15th, 1864.

"*Dear Madam*—I enclose you a lock of hair which I clipped from Phares' head. I suppose you have heard some of the particulars of his death. It was so sudden to us all, that it hardly seemed like death. An hour before, he sat in my tent with me, chatting of the times when we were boys playing ' Robin Hood,' and roving through Bailey's Grove—of school days—then of our late capture, and its strangeness, compared with those times. After awhile, he arose, saying, that he thought he was going to have a chill, as he felt very much like it. Twenty minutes after, one of the boys came rushing into the tent, saying that Phares was dying. I hastened to his tent only to find him dead. I had a surgeon called immediately, who pronounced his ailment—' heart disease.' If I am ever fortunate enough to reach Perry, I will tell you all.

Yours truly,

"J. W. M."

183. SHOCKENSEY, TIMOTHY F., China.—Enlisted August 30th, 1862.

Was taken prisoner at Plymouth, and died at Andersonville Stockade, September 12th, 1864. The number of his grave is 8,595.

He left a wife and family to mourn his loss.

The sympathy of all who appreciate the sacrifice that a man with a family made, when he left his home to enter the ranks of our army, as well as the kindest wishes of all his comrades, is tendered to them.

184. SMITH, MASON C., Perry.—Enlisted August 28th, 1862.

Mustered in at Buffalo, August 30th, 1862.

Was appointed corporal at Newport Barracks.

Was taken prisoner at Plymouth. The writer saw but little of him until we reached Andersonville. I was then called upon by one of the "battery boys," who informed me that Mason was very ill, and desired to see me. Upon visiting him I found him very low, with unmistakable symptoms of typhoid fever. He recognised me for a few moments only, spoke of home, his mother, and of our old boy days, and then wandered off in mind to the insane dreams of a fevered brain. Even then, as we gathered about him, it appeared to us that to be allowed to be the first one to die in such a place, was a kindly boon of a wise God. He died the following morning, May 10th, 1864. The following obituary appeared in the *Western New Yorker*, when the sad tidings of his sad death reached his home:

OBITUARY BY REV. J. R. PAGE.

"Another noble sacrifice for our imperiled country. Last Saturday, a letter was received from Lieut. George S. Hastings, who is a prisoner at Macon, Georgia, containing the brief sad line—'Mason Smith died, May 10th.'

"This much, and no more! His health was quite poor at the time of his capture, and he, doubtless, sank under the long, weary march, and the increased hardships incidental to his captivity. Our community could mourn the loss of no young man more beloved or promising.

"He inherited the genial, quiet, kind spirit of his father; had the same relish for literary pursuits and social enjoyments; was equally intelligent and uncompromising in his convictions, and bid fair to be a man of even greater usefulness and worth to society. He had been for several years an earnest, active member of the Presbyterian Church, deeply interested in the Sabbath school and prayer meetings; honored by all his youthful associates for rare and moral courage, and manly adherence to what he regarded as the path of duty. For, rather would he right, all alone, than wrong, with the multitude.

" He had just completed his second year at Hamilton College when the call for troops became too earnest for him to longer resist. Had he remained in college it is quite likely he would have returned to his home from the recent commencement, a graduate, in the very stage which brought the intelligence of his death. * * *

"The frequent letters received from him showed that he aimed to meet all his obligations, and maintain his integrity amid abounding temptations. His efforts were more successful in the opinion of others than in his own severer judgment. Very characteristic is the following extract from his last letter, save one—' In regard to my being corporal, I have only to say, it was nothing of my own seeking; I never curry favor of my officers—I simply do my duty—and what they see fit to give me, I take—even if it is no more than a corporalcy. Non-Commissioned Officers in the Twenty-fourth Battery are very precarious. To day you may be a sergeant, to-morrow a private. The company is full of men who once held a position in the battery.'" * *

The following appears in an obituary, written by the secretary of his college class:

"As a class-mate, he was loved by us all. Few enjoyed the popularity that he was held in by all the class. Quiet, but determined, he was first in our sports and plans; and by his genial spirit, made all his friends.

"A true Christian and an earnest worker; we can but mourn his sad fate, and join our grief with that of his bereaved family in the loss to them of an only son and brother; to us, of an honored and respected classmate."

185. SMITH, J. W., Kingston.—Enlisted September 29th, 1864.

Transferred to Third New York Artillery, May 25th, 1865.

Joined at Roanoke, October 17th, 1864.

186. STEVENS, GEORGE W., Fort Plain, N. Y.—Enlisted November 24th, 1861.

Re-enlisted in January, 1864, at Plymouth.

Taken prisoner at Plymouth, and sent to Andersonville.

Phelan informs us that he died at Florence.

187. STODDARD, SAMUEL.—He writes: "I enlisted September 5th, 1862, at Perry, N. Y. Was mustered in, September 10th, 1862, at Buffalo. Mustered out at Syracuse, N. Y., July 7th, 1865.

"I was not a prisoner. I barely escaped capture at Plymouth, N. C., April 20th, 1864. It happened in this wise: I left Plymouth in company with Sergt. Camp, April 3d, for Washington, D. C. Having finished our business, we received orders to return on the 14th. We left Washington the same day, expecting to reach Plymouth on the night of the 16th, or on the following morning, but owing to the failure of connection at Norfolk, of about an hour, with the Chesapeake and Albemarle Line connecting Norfolk with Roanoke Island, we were detained at Norfolk until the following Monday, April 18th, the day after Plymouth was attacked.

"We were joined at Norfolk by Tom McGuire and Parmlee, who had been left behind at that point, sick, on their return from their veteran furlough.

"On reaching Roanoke Island, we found that we were just one hour too late to reach Plymouth, as the ram came down the river that night, and cut off all further communication with the place from our transports. We, however, were ignorant of this, and proceeded by the 'Massasoit,' to join the Battery at Plymouth. When about midway of the Sound, however, we hailed one of the gunboats, having on board the body of Capt. Flusser, and learned the situation of the garrison. We steamed on, notwithstanding, and joined our fleet, now lying in Chowan Bay, where we remained all night. On the morning of the 30th, we steamed up to the mouth of the Roanoke, and there, during the day, helped off refugees and some escaped prisoners who had found their way to that point.

"At night we were transferred to a propeller that had been trading in those waters, together with those picked up day by day, and several enlisted men that had arrived from New Berne that day, belonging to the Battery. Were sent back to Roanoke Island, then under command of Lieut.-Col. Clarke, Eighty-fifth New York Volunteers.

"On reporting to Col. Clarke, Sergt. Camp was detailed to the quartermaster's department, where he remained until he received his commission, in the following February, I think. This left me in command of our squad, which consisted of the two veterans above named and five recruits, and with them I was ordered to report to Capt. Barnum, of the Sixteenth Connecticut Volunteers, stationed at Fort Reno.

"Here I remained for nearly a week, when I was detailed to report at headquarters, as clerk to the acting assistant adjutant general, and in which place I remained until about the first of June, 1865. During these months I had the opportunity of rendering some assistance to those self-denying ladies who were sent as teachers to the contrabands, as it was my privilege to do to some extent at Plymouth. I will mention another incident, which was of interest to me—a thing which comparatively few saw while in the army—*a revival of religion*, in which many were hopefully converted, both officers and enlisted men.

"I shall not soon forget the testimony of one of the Andersonville prisoners, who had returned for duty with his regiment. In speaking of the change he experiences in becoming a Christian, he said in words, as near as I can recall them: 'I have been, as you know, for some months a prisoner, in the hands of the Rebels at Andersonville, and I thought while descending the river to the point our exchange boat was stationed at, as I first caught sight of the old Stars and Stripes, that it was the happiest moment of my life; but I can assure you that this comparison but feebly expresses the joy which I now feel in becoming a child of God.'

"I was appointed corporal, October 11th, 1862, and

was mustered out as such at the expiration of my term of service.

"Since my retirement from the service I have resided in New York City for three years, as a student of Theology, at Union Theological Seminary, spending my vacations, however, out of the city. My first vacation, during the Summer of 1866, was spent as an agent of the Freedman's Union Commission, and canvassed Rockland Co., N. Y., collecting funds in behalf of that cause. During the vacation of 1867, I preached as stated supply of the Presbyterian Church at Stanhope, N. J., having been licensed to preach the Gospel, April 13th, 1867.

"On May 9th, 1868, graduated at Union Theological Seminary, and on the 20th of the same month was married to Miss Sarah E. Hoisington, daughter of the late Rev. H. R. Hoisington, many years a missionary of the American Board of Foreign Missions, and for several years principal of the Batticolla Seminary, Ceylon, where Mrs. Stoddard was born.

"In June, 1868, received a commission to labor as a home missionary, under the auspices of the Presbyterian Home Mission Committee. I was sent to Holton, Kansas, which is now my field of labor, and present address."

188. STORMS, THOMAS S.—Enlisted at Tarrytown, November 6th, 1861. Was in the Battery. Discharged, we believe, on account of some physical inability, and is now living in Tarrytown.

189. SUNDERLAND, CHAS.—Enlisted at Rochester, September 20th, 1864. Joined the Battery, November 9th, 1864, at Roanoke.

He enlisted for the 108th Regiment Infantry, but was "lost in the wilderness," and the Battery claimed him.

190. SUNFIELD, JAMES, Rochester, N. Y.—Enlisted October 9th, 1861. Re-enlisted as a veteran, at Plymouth, in January, 1864.

Was taken prisoner at Plymouth, and sent to Andersonville. He was one of those fortunate few who miraculously escaped from the jaws of death.

He was reported as "absent at College Green Barracks, Annapolis." We have been told that he now lives in Rochester, N. Y.

191. THAYER, LEWIS P.—Enlisted at Rochester, October 4th, 1864, for one year. Joined at Roanoke, November 9th, 1864. Transferred to Third New York Artillery.

192. TILTON, HENRY, Moscow.—Enlisted August 29th, 1862.

Mustered in at Buffalo, August 30th, 1862.

Promoted corporal about October, 1863.

Was taken prisoner at Plymouth, and died at Andersonville Hospital, of gangrene, October 18th, 1864.

Tilton was one of three brothers who entered into the service of the United States. We believe he was the second one to go, and the second one to die. His brother had told him of severe war experience, but that did not deter him. When in his father's store his associates and he consulted over the proposition to go, and go together, he was one of the strongest advocates of the plan. It required much persuasion to gain his mother's assent (for

WASHING AWAY THE ANDERSONVILLE STOCKADE.

she could realize, perhaps far better than he, the possible sufferings that he might endure), yet, full of enthusiasm, and full of an earnest conviction that he was needed, he did gain it. And we know, too, that another dear one protested and pleaded; but the firm resolve of a convicted mind gained the mastery over the heart, and he bade them all a "good bye," satisfied that he had done right. An impression has been given to his friends that his sufferings from the disease of which he died, were extremely painful and severe. This is not true. The writer knows better than anybody else, because he constantly attended him, and had different physicians to visit him. Scorbutis made its appearance in his face. While he remained in the stockade, this disorder received little or no attention. A large ulcer formed in the cheek. He affirmed that he had no sensation of pain from it, and that it was callous to the touch. When he reached the hospital and found friends, was decently clothed, comfortably sheltered, and had received some palatable food, he volunteered the remark that " He should soon get well, now that he could get something to eat."

He had been in the hospital once before, and had been returned to the stockade as well. We believe that the cause of his death was not gangrene, but debility, arising from want of food and want of shelter, before it was too late. He received all the comforts that his several friends were enabled to give, for all of which he was very grateful. We hoped to save him, but he was too far gone. Upon making the usual morning visit to his tent, after he had been there a few days, we found that his soul had passed quietly away during the night.

The gracious God, who had given him rest in slumber, had carried him from that sleep to the eternal sleep which knows no waking. Will there not be rest in heaven for such a wearied, suffering martyr?

193. TIRRELL, SAMUEL.—Enlisted at Leicester, January 5th, 1864. Reached Plymouth in time to participate in the battle, and be taken prisoner.

Ferguson writes that "Tirrell died at Florence, S. C."

194. TRUAIR, O. M., Mount Morris.—Enlisted September 4th, 1862.

Quaker. Died or was discharged.

195. TURNER, ROBERT, New Hartford, N. Y.—Enlisted November 22d, 1862.

He came on to Newport Barracks with Lieut. Hastings, having left Hamilton College in order to enter the service.

He was unusually bright and active, impulsively generous and kind, and very popular with the members of the Battery. So anxious was he to participate in a battle, that he used his strongest persuasive qualities with the officers to permit him to go with the section which was selected for the march to Kinston, Goldsboro' and Whitehall. He was killed at the battle of Whitehall.

The following account is given in "The Wyoming Times":

"Monday morning the army re-crossed the bridge, it was a long, large bridge. Robert Turner, from Owego, had charge of men detached from the different batteries, to clear the town of stragglers, and send them forward to prepare material, and make preparations

for burning the bridge and set fire to it. He accomplished the deed promptly, and received the approval of the Colonel. * * *
(We believe this was the bridge on which Colonel Clark, of the Ninety-sixth New York, was shot and killed.) At the battle of Whitehall, Turner was standing at the head of a horse, when he was struck by a fragment of a shell, killing him instantly. The missile passed through his body near the heart. He was buried in his overcoat and blanket, even in the din and smoke of battle.

"He was a great favorite with all the men, and sad hearts gathered round his lonely grave. A short prayer was said, and there was just time to fill the grave as the command was given to 'forward.'"

196. VAN BUREN, SYLVESTER.—Enlisted February 15th, 1864.

He was taken prisoner at Plymouth, and was sent to Andersonville.

197. WARDWELL, EDWARD H., New Hartford, N. Y.—Joined for duty, September 20th, 1862. Promoted second lieutenant, April 15th, 1863.

Resigned, August 30th, 1864.

Transferred to Signal Corps and absent from Plymouth.

198. WASHINGTON, GEORGE, colored cook.—Enlisted at Plymouth, May 11th, 1863. Lieutenant Camp says: "He was taken prisoner at Plymouth, April 20th, 1864. Put under guard and set at work collecting stores and carrying them into Fort William.

"While at work, the third day after capture, he gave the guard the slip, by jumping over the parapet into the ditch, near our quarters, thence into the swamp, on the upper side of the town, where he found a canoe, with a gun in it, which some person had left and gone ashore from. He got into it, paddled out into the creek, thence to the river, thence up the river, past the Fort and down Middle river, picking up in the swamp, opposite Plymouth, two refugees, N. Carolinians, and brought up safe and sound with our fleet, at the mouth of the river. Joined the company at Roanoke, and for some misdemeanor was sent for trial to New Berne, there got out and went on board steamer as coal heaver. From George, Lieutenant Camp got the first report of the killed and wounded, and fate of the Battery, in general, which he conveyed by letter to anxious friends at Perry."

199. WAYNE, JOSEPH, Hamlin, N. Y.—Joined for duty October 23d, 1861.
Discharged from hospital in June, 1862.

200. WELCH, EDWARD, Perry.—Enlisted, August 27th, 1862.
Mustered in at Buffalo, August 30th, 1862.
Was wounded and taken prisoner at Plymouth, and died at Andersonville Hospital, of chronic diarrhœa, August 8th, 1864.
The number of his grave is 5,181.
Welch was one of the hardiest men in the Battery. Once or twice before he was taken prisoner, he had been attacked with fever and ague, but under any ordinary endurance he doubtless would have survived many others.
His death was quiet, and somewhat unexpected to us all.

201. WELLER, JACOB H., Cuylerville.—Enlisted August 28th, 1862.

Mustered in at Buffalo, August 30th, 1862.

Taken prisoner at Plymouth, April 20th, 1864, and was sent to Andersonville.

202. WETMORE, CHAUNCEY, Hague.—Joined for duty, September 28th, 1861.

Re-enlisted at Plymouth, in January, 1864, and was taken prisoner. He was at Andersonville some time, and was finally removed to Florence, where he died.

Phelan thinks that he died at Andersonville.

203. WHITNEY, HAMILTON S., Albany.—Joined for duty, November 16th, 1861.

Re-enlisted at Plymouth. Was married to Elizabeth Owens, while on his veteran furlough. Taken prisoner at Plymouth. He lived through the imprisonment and was exchanged, December 15th, 1864. He was mustered out with Ferguson, W. Carnahan, and Holman, at Syracuse, July 7th, 1865.

Present address, Johnstown, N. Y.

204. WHITNEY, W. A., Wellsville.—Joined for duty, November 21st, 1861.

Re-enlisted in January, 1864, at Plymouth.

Appointed bugler. Transferred to Third New York Artillery, May 25th, 1864.

205. WHITBECK, HENRY, Hamlin.—Joined for duty, October 23d, 1861.

He died in January, 1862, in Washington, of the measles. It was the first death in the organization.

206. WILLIAMS, OLIVER, Perry.—Enlisted August 26th, 1862.

Mustered in at Buffalo, August 30th, 1862.
Promoted corporal in December, 1862.
Promoted sergeant.
For a short time acted as orderly sergeant.
Was in command of a section of the artillery at the battle of Plymouth. Was taken prisoner at Plymouth, and died at Andersonville Hospital, of intermittent fever, July 24th, 1862. The number of his grave is 3,947.

Williams' enlistment in the Battery was not his first attempt to add his name to those who responded so promptly to the call of their country; so thoroughly was he convinced of his duty, so firmly fixed in his determination to go, that neither argument, persuasion, nor tears, availed aught against his decision. Dear were the ties that bound him to his home—bright were his future prospects, should he remain; but what were these, when his duty was clearly demonstrated to his mind, and he saw those who had been his friends and associates from his early childhood, gathering around the standard.

He proved to be a man of worth, in the vocation of a soldier—he was fearless and proficient—his duties were promptly and satisfactorily completed—and as he showed his worth, he was promoted.

His section at the battle of Plymouth did good service, although they were among the first that were captured, or rather overwhelmed. He was sent to Andersonville Hospital from the Stockade, with a squad of sick men; his friends found him, and from that time until his death, he was well cared for.

We are aware, from conversations with him a few days prior to his death, that he felt that he would not recover. We believed he would, and we encouraged him to think so; but intermittent fever is an insiduous disease, and its victim, one day seemingly bright and improving, passes away the next, to the amazement of even the physicians themselves. We believe that Williams was prepared to die, and was resigned to the will of his Heavenly Father.

207. WILLIAMS, THOMAS.—Enlisted October 4th, 1864, at Rochester. Joined at Roanoke Island, November 9th, 1864. Transferred to Third New York Artillery.

208. WINNE, BARNETT V. L., Albany, N. Y.—Enlisted September 7th, 1864. Transferred to Third New York Artillery, May 25th, 1865. Joined for duty at Roanoke, October 17th, 1864.

209. WOOD, EMMETT, Moscow.—Mustered in at Buffalo, August 30th, 1862. Taken prisoner at Plymouth, and died at Andersonville Stockade, of chronic diarrhœa (so said), September 1st, 1864. The number of his grave is 7,581.

With others, Wood at one time endeavored to make his escape. They succeeded in getting away from the hospital, but the flight was discovered in the morning, and the dogs were put on the track. The hounds overtook the party, and Wood had a portion of his ear torn off by them.

He recovered from this after he was brought back, but he was placed in the stockade, and of course he was a marked man after this; consequently he undoubtedly was denied many things that were furnished to others.

210. WOOLSEY, JOHN.—He writes: "I enlisted in New York City, November 18th, 1861. Was mustered in about the 1st of December, 1861.

"We were to be armed with 'Congreve Rockets,' and were led to believe they were a very effective weapon. Our commander, Major Lion (in a speech he made to us at Albany), told us that they were used with great effect in the Mexican War, one going a mile out of its way to kill a Mexican. (I have no doubt it struck a mile from the object it was aimed at.) Early in December we were ordered to Washington, where we were encamped about a mile from the Capitol. There we received the long expected 'Rocket Guns.'

"We took them over to the east branch of the Potomac to try them; we expected to see wonders (as we had been told that, with a little practice, a flagstaff could be hit five miles off). An artillery blanket was hung up for a target, about three-fourths of a mile distant. While we were shooting at it, some cold-blooded scamp stole the blanket. The rockets were balkey; like a mule, they would go any way but the right way. That night rocket stock was low in camp. The next day they were returned to the armory, and we received three inch rifled guns.

"One of our officers is worthy of mention: during that winter he used to appear on the parade ground at the morning drill with his head in a woolen tippet, his pants in his socks, and his toes in a pair of slippers. He would watch the drill, and seeing something wrong, he would rip out a volley of orders, at the same time pointing a clay pipe at us in such a manner, that if Barnum or Dan Rice could have seen and heard him, his fortune would have been made. The name of Rathbone was frequently heard to echo through the camp.

"At the battle of Plymouth, while loading the gun, in order to give the near approaching rebels a last shot, I was hit in the right thigh by a fragment of shell, and a few seconds after was shot through the right arm with a musket ball, breaking it so badly that a part of the lower bone had to be removed, which makes me partly a cripple, and I now receive a pension of eight dollars per month.

"I remained a prisoner at Plymouth about two weeks, when I was sent to Weldon by boat. I was then too weak from fever and inflammation to stir (except my tongue). I was put in a mule wagon, with a darkey on one mule, and started for the Raleigh depot. We had to cross about a dozen switches quartering. I repeatedly invited the darkey, in very strong terms, to drive slowly, but the more I urged him, the louder he sung to the mules, 'Get up there, what I feed you for!' He undoubtedly enjoyed that ride better, and will forget it sooner than I shall. At Raleigh I had the erysipelas in my arm, and should probably have died but for the attention of friends. In June I was sent to Salisbury; there were not many prisoners until October, when about 11,000 were sent there. The scenes of that Winter you are probably familiar with. I there met a man named Ainsworth, a brother to Rufus and William. I do not know whether he lived to get home or not.

"On February 22d, 1865, I started for Wilmington to be paroled. Walked from Salisbury to Greensborough (fifty miles), on the railroad track; signed my parole at Goldsborough, March 1st, and marched into our lines at Wilmington, the 2d. Was sent to Parole Camp at Annapolis, where I was sick in the Hospital for a time, When I entered the hospital I put $80 (that I had just received as ration money) in a safe that was kept for the purpose by the surgeon. When I went after it I was told that one of the clerks had run away with $5,000, mine among the rest. I have been unable to learn whether he ever stopped running or not. I did not reenlist, and my time had been out since November, and I was ordered to Albany, to be discharged. I proceeded there, and after waiting a month, was mustered out May 3d, 1865. Since that time I have been on a farm in Westchester County, New York, until last Fall, when I came to Iowa, and have been teaching school in Crescent, Pottawottomie County, this Winter; but I intend to return to New York in a couple of weeks. I have thus far been unable to find any one that is willing to be the wife of a crippled soldier."

Present address, Bedford, Westchester County, N. Y.

211. WOOLSEY, ETTING, Albany, N. Y.—Enlisted September 7th, 1864. Transferred to Third New York Artillery, May 25th, 1865. Joined at Roanoke, October 1st, 1864.

212. WRIGHT, GEORGE G., Hamlin, N. Y.—Joined for duty, November 9th, 1861. Re-enlisted at Plymouth, January 1st, 1864.

Was promoted corporal, and retained this position until he was mustered out. He was taken prisoner and sent to Andersonville. Was paroled for exchange in November, 1864. Joined the Battery again for duty, April 12th, 1865. Was mustered out with the remnant of the Battery, in Company " L," of the Third New York Artillery.

We have heard that his present address is Rochester, N. Y.

213. YANCER, J. D.—Joined for duty, February 8th, 1864. Reached Plymouth in time to be taken prisoner, and was sent to Andersonville. Mosier writes that—" Yancer died at Andersonville, August 15th, 1864."

Part II.

CHAPTER I.

1861.

THE ROCKET BATTALION.

In the local columns of the Wyoming *Times*, under date of September 27th, 1861, we find the following paragraph :

> Meetings and speeches in favor of the war we had supposed "played out." *Action*, ACTION, is now the word. All are enlightened on the subject of the war, or ought to be.
>
> Monday evening, however, another meeting was held, called by Messrs. Wyckoff, Lee and Page, with a view of obtaining recruits for a company of artillery.
>
> Prof. Atkins was called to the chair ; whereupon Jay E. Lee Esq., stated it was their purpose to organize an artillery company to be attached to G. D. Bailey's regiment, and enlarged at some length upon the advantage of this branch of the service over all others.
>
> He was followed by Harry C. Page, Prof. Atkins, Rev. Mr. Tomlinson, Rev. Joseph R. Page, Judge Gilman, N. P. Currier and Philander Simmons ; after which an opportunity was given to enlist.

The result of this meeting was a response from about fifty men to the call, who pledged themselves to the organization proposed. For some reason, which we are unable to explain, only twenty of this number kept their faith. A correspondent, over the *nomme de plume* of "Drummer," writes to the *Times* under date of October 29th, 1861:

> Most of your readers are perhaps aware that some fifty individuals signed their names to the roll at Perry, and were expected to go with the company, in addition to those whom we expected to join us from Monroe County. Well, on the morning of our departure we could find but twenty persons prepared to go. Nevertheless, according to previous announcement, we took the cars at Castile last Friday morning for Buffalo, with the understanding, if we found our Monroe friends on hand and a fair prospect of filling up our company soon, we would be sworn in and organize; otherwise, we should return.
>
> At Buffalo we met twenty hearty and determined men, and, after looking the ground over, were mustered in. Every man who went with us, except Wyckoff, was sworn in.
>
> * * * * * * *
>
> We were sorry to lose Mr. Wyckoff, but, as we had so few men, we could not ask for both captain and first lieutenant; Mr. Wyckoff, therefore, though urged to remain, reluctantly withdrew. "If I cannot bring more than twenty men," said he, "and not half those men care whether I go or not, I shall not stay." That he did not go with us as captain is the fault of those who signed and failed to go. If we had gone to Buffalo with forty men, Mr. W. would have been captain.

Having completed the organization, the company remained at the recruiting head-quarters—Fort Porter, Buffalo—until about the middle of November. They then left for Albany with 56 men. While at this post, Major Thomas W. Lion, ex-English army officer, inventor of the wonderful fire-rocket, &c., &c., introduced himself to their notice. He desired to form a battalion, to use this *rocket* in the field. A consolidation of several squads of recruits, occupying the barracks at Albany, then formed

THE ROCKET BATTALION.
Major—Thomas W. Lion.

Company A.
Captain—A. Ransom.
1st. Lieut.—H. W. Dodge.
2d Lieut.—Samuel Hoddy, Jr.

Company B.
Captain—Jay E. Lee.
1st Lieut.—L. A. Cady.
2d Lieut.—G. W. Graham.

The battalion consisted of 160 men, equally divided between the two companies.

We are, however, only interested in the history of Company B. The non-commissioned officers appointed at that time in this company were:

Mark Andrews, Orderly Serg't.	H. C. Page, Q. M. Sergeant.
Wm. W. Crooker, 1st Duty Serg.	Rufus Ainsworth, 3d Duty Serg.
Solon Rowell, 2d " "	Robert Bullock, 4th " "
Charles A. Clark, 1st Corporal.	Gustavus Barker, 5th Corporal.
Chas. R. Griffith, 2d "	L. J. Sanford, 6th "
Franklin D. Otis, 3d "	G. W. Kellogg, 7th "
Geo. B. Johnson, 4th "	Francis Leonard, 8th "
Hector C. Martin, Bugler.	L. Newcomb, Bugler.

Albert Richards, Artificer.

In December, the battalion received orders to report in Washington. Leaving Albany, December 7th, they proceeded on the steamer *New World* down the Hudson to New York. Were detained a day or two at the Park Barracks at New York City, and departed thence by railroad to Washington. There they went into camp on the artillery grounds, east of the Capitol. There were few incidents connected with this trip that would be of sufficient interest to relate. Just before reaching New York a fog enveloped them, and the steamer ran aground, and they were delayed some time. A discovery of some country produce among the freight packages of the steamer, furnished the boys with a dessert for their haversack dinner. The short stay in New York gave them an opportunity to do some sight seeing, and we judge that one of their number had "seen" a good deal, when he stepped into a millinery store and told the smiling young lady attendant, with his most winning manner, that he "thought he would take a whiskey sour." The young lady assured him that they kept nothing sour, but he would doubtless find it next door. "By George, I thought this was the right church, but I guess I am in the wrong pew," was his remark to himself in *sotto voce*, as he made a hasty exit.

The Wyoming *Times*' regular correspondent, "Quartermaster" (Harry C. Page), writes under date of January 6, 1862:

* * * * * * * *

I said we left Albany, December 7th. In New York we were entertained at the Park Barracks, where we all slept for the first time, I doubt not, with the "gates open." A more vile and miserable den men cannot be packed into. It is the "Elysium of Loaferdom." I wonder if there will be any improvement under the new Mayor. Fortunately our stay in New York was brief. [The condition of Park Barracks was directly the reverse of this description in the Fall of 1862.—ED.] On the evening of the 9th, we took the cars for Philadelphia, where, at midnight, we partook of a splendid repast, prepared by the ladies of that hospitable city, who, we were informed, had in like manner appeased the appetites of 100,000 soldiers since the war began. Upon the arrival of a train bringing troops a gun is fired, and whatever be the hour, day or night, a committee of ladies repair from their dwellings to the room prepared for that purpose, to serve out warm coffee and other food to the soldiers. There we took the cars for Baltimore, and few of us, as we marched through the streets of that city, but thought, I fancy, how different was our reception from that of the

Sixth Massachusetts Regiment, which, on the 19th of April last, on its way to defend the capital of the nation, was assaulted by an infuriated and misguided populace.

* * * * * * * *

We reached Washington, December 10th.

* * * * * * * *

We copy from the Brockport *Republican* a brief sketch, from the pen of Captain Lee, of this portion of their trip :

The contrast between our welcome in Baltimore and that of some of the first troops that passed through there was most striking ; indeed, our journey all through Maryland was a complete ovation. The women and children, of all classes and descriptions, hurrahed, waved flags, handkerchiefs and petticoats. A few in Baltimore looked savage and muttered.

It seemed strange to me that the citizens of Philadelphia and Baltimore should pay us so much attention. They greeted us as warmly and entertained us as generously, as though we were the first soldiers who had passed through their cities, and as though we were the sole saviors of the country.

CHAPTER II.

1862.

THE ROCKET.

It was a "fizzle"—yet it was an invention, or claimed to be one. It had a man of brass in Major Lion, to place it before the Secretary of War and the Chief of Artillery—so plausible was the theory, that the expense of sustaining the Battalion, as well as many other expenses were incurred, to give it a practical test. In the editorial of the Wyoming *Times*, December 20, 1861, we find the following, probably from the pen of T. S. Gillet, who was then in the office of Secretary of State, at Albany:

> Various statements have appeared in the papers relative to the "Rocket Gun," and none precisely alike, yet all representing this arm of the service as a most terrible one. It has never been used on this Continent, and experienced artillerists have never seen it. The papers and Government are only in the secret. Its principal purpose appears to be to throw forward a flame of fire sufficiently large to frighten horses and thus throw the enemy's cavalry into confusion. Of course, the battalion must have the right of the advancing army, and take their chances of having their 'Rockets' silenced by the picked riflemen of the opposing forces.
>
> The "rocket" gun is represented as being a breech-loading field piece, capable of discharging bombs, balls and percussion shot as well as rockets. The rockets are to be used for firing buildings, behind which

the enemy may seek shelter, or for removing by fire any obstacle thrown out to retard the advancement of the troops. The expansive properties of the rocket are wonderful, creating a ball of fire fifteen feet in diameter, which can be thrown by this breech-loading projectile 5,300 yards, or over three miles!

It is stated that the Government has purchased the exclusive right of manufacturing this terrible instrument of destruction, and is soon to introduce it to the rebels.

"Quartermaster" furnishes the Wyoming *Times*, of date January 24th, 1862, a more elaborate description of these instruments of war. He says:

I will give you an imperfect description of the missile and its use—

The rockets which I have seen, vary from twelve to twenty inches in length, and from two to three inches in diameter. The head is conical and solid iron, from two to three inches in length, according to the size of the rocket. The remaining portion of the rocket is a hollow iron tube, filled with a highly inflammable compound, which being ignited in the rear or tail of the rocket by a fuse, gives the weapon its impetus.

The composition of this inflammable substance is a Government secret.

To form some idea of the noise and force which they made when fired, you may multiply the noise and fury of a large Fourth of July rocket by one hundred. We have made but one experiment with them since coming here, and that was at the arsenal, and more for the purpose of testing some conductors or tubes from which to fire them, than the rockets themselves.

The tubes we used were of two patterns, one of drawn iron with a bore of three inches, and the other by uniting three three-quarter inch rods or wires, spirally, fastened by strong collars or bands, leaving a bore or tunnel of about four or five inches. Both were placed on a stand somewhat similar to a theodilite stand. The rockets used were old and not very perfect, yet we executed some very satisfactory firing. The results from the wire tubes were most satisfactory. Two three-inch rockets fired from the latter, went magnificently. The tube was pointed across the river (Potomac), diagonally, at an elevation of nearly 45°. Away went the fire spitter, out of sight, and probably found a grave in the " sacred soil." Turning the tube down the river, at the same elevation, a second was fired, it went beautifully —direct as the path of a bullet—and buried itself in the Potomac, at a distance of more than three miles. That was the estimate of General Barry, Major Ramsey, and others familiar with the locality. The rockets we are to have for active service are a decided improvement on those we used, which I have described.

The head, instead of being solid, will be hollow and filled with musket balls and powder and exploded by a time fuse, in all respects similar to a "grapnel" or "spherical case " shot. The advantages from that improvement are palpable. The head will be heavier (on account of being filled with lead,) which will materially add to the directness of the line of flight and to the distance. Then by being fired by a time fuse, it can be exploded at any time or place, scattering a storm of bullets and fragments around. Another improvement is this : the tubes or case containing the combustible material is to be perforated by tangential, spiral holes, from which the fire will be thrown with great force and fury, giving a whirling motion to the missile, which as you see, will also assist in giving directness and distance to its flight as well as scattering fire and destruction on every side.

Our organization is the same as light artillery. We shall have gun carriages and limbers, followed by caissons. But instead of mounting one gun on a carriage, we shall mount four rocket tubes. Our company will work four carriages and its guns or tubes. Just think of us drawn up in battery before a regiment of cavalry or infantry. At one volley we could send into their midst sixteen rockets, each rocket spitting fire, fury and destruction on every side, and carrying in its forehead seventy-four bullets, ready to burst from their shell at just the desired point, and scatter death in every direction.

If all this succeeds in the field, as it is believed it will, our weapon will be terrible in its execution, and we a terror to traitors.

The " Rocket guns," after a long delay of nearly four months, were turned over from the hands of the inventor and contractors to the battalion. " Quartermaster " writes, March 31, 1862 :

We have been encamped here nearly four months, and have just got our guns. The principal cause of the delay arose from the fact of the guns being of a peculiar construction, and we were therefore obliged to wait for their manufacture. They have at length arrived.

The carriages are lighter than those of light artillery. The tube is made of wrought iron, and is eight feet in length, with a bore of two and a quarter inches. The tube is perforated with holes about one inch in diameter, the entire length, and about two inches apart.

The object of the holes is to permit the flame to escape while the rocket is passing through the tube, which might otherwise be corroded. We are immediately to commence experimenting with the guns. * * Nothing can much longer delay us, unless it is the scarcity of horses.

A week later, he writes :

I wrote you, one week ago, that we had received our rocket guns. Since then we have had our horses. Recent experiments with the rocket rendered certain their perfect success and immense power as a weapon of warfare, so you may soon expect to hear from us on the field of battle.

May 12th, 1862, he writes, in transports off New Berne :

We made some experiments with the rockets shortly before we left Washington. They hardly came up to the expectations of the authorities, and so it was concluded, as Burnside was in want of artillery, to give us some guns and send us on. Consequently our quaint rocket carriages were exchanged for the substantial six-pounder carriage, and our sheet ironed tubes were turned into rifled cannon.

This closes the correspondence on that interesting subject, the "rocket gun."

The writer has had many conversations with the officers and members of the Battalion who were present at the trial of these guns.

The fault which proved too great to overcome, was

the uncertainty of direction that the rocket would take. It might make a retrograde movement. It might, immediately on its leaving the mouth of the tube, take a counter direction and come flying into the midst of those who fired it. A body moving from the centre of a circle, there was no knowledge of the radius it would probably take in its flight.

The rocket used was, by Major Lions' representation, an improvement, invented by Lion himself, upon the Congreve rocket.

In the minds of those best acquainted with him, I find that there were doubts as to whether he knew anything at all about the science of gunnery or of this projectile.

In the text book on "Ordnance and Gunnery," by Colonel J. G. Benton, used in the U. S. Military Academy, this short history of Rockets is given :

Rockets were used in India and China, for war purposes, before the discovery of gunpowder ; some writers fix the date of their invention about the close of the ninth century. Their inferior force and accuracy limited the sphere of their operations to incendiary purposes, until the year 1804, when Sir William Congreve turned his attention to their improvement. This officer substituted sheet iron cases for those made of paper. which enabled him to use more powerful composition ; he made the guide stick shorter and lighter, and removed a source of inaccuracy of flight by attaching the stick to the centre of the base instead of the side of the case. He states that he was enabled by his improvements to increase the range of six-pounder rockets from six hundred to two thousand yards. Under his direction they were prepared and used successfully at the seige of Boulogne, and the battle of Leipsic. At the latter place they were served by a special corps.

The advantages claimed for rockets over cannon are unlimited size of projectile, portability, freedom from recoil, rapidity of discharge, and the terror which their noise and fiery trail produces on mounted troops.

The numerous conditions to be fulfilled in their construction, in order to obtain accuracy of flight and the uncertainty of preserving the composition uninjured for a length of time are difficulties not yet entirely overcome and which have much restricted their usefulness for general military purposes.

From this description we are led to conclude that there was some ground for the representations of Major Lion, and good reasons for the high expectations which the officers and men of the Battalion had in a probability of becoming "the pioneer organization of this wonderful arm of the service."

The failure did not certainly arise from want of patriotism, courage, or willingness on the part of the members of the Battalion.

CHAPTER III.
1862.

BATTERY B.

Six months had elapsed while these experiments on rockets were being made.

It is easy to imagine that the days dragged slowly by with those who had, at the time of their enlistment, expected immediate active duty.

Six months of settled camp life, where the daily routine is one of leisure, is very likely to be demoralizing to any company.

Aside from a little foot drill and sabre exercise, their time was unoccupied by regular duties. An occasional show of discipline was made by the commander of the Battalion. This depended, perhaps, more on his mental and physical condition, than upon any direct or flagrant misdemeanor.

Sectional and personal jealousies arose among the men and officers. Against the officers charges of injustice and favoritism were made. Counter-charges were made by the officers of inefficiency and insubordination.

Their troubles and differences were brought before the court-martial, and both there and at the homes of those interested, the various actions in the matter were thoroughly discussed.

We do not propose to open the discussion again, and mention this portion of the Company's history, only because they were facts and incidents occurring at that time.

The friends at home had not forgotten the volunteers, and we have accounts of feast and joy over "good things" from home.

There was some sickness and one death—that of Lemuel Andrus.

In order to give a better understanding of the record of their life at Washington, we select from the files of the Wyoming *Times* the following correspondence:

CAMP CONGREVE, December, 1861.

You will see by my date that the name of our camp has been changed from Duncan to Congreve, in honor of the inventor of the rocket. It is located about three-quarters of a mile from the Capitol, on what would be the prolongation of Pennsylvania Avenue, if that thoroughfare was continued through the Capitol. Let me describe my quarters to you. My house is what is called a wall tent. The boys have planted a row of young spruce trees in front of the tent, which is quite ornamental. It is heated by a California stove, the institution of the establishment. This consists of a hole near the centre of the tent, about eighteen inches deep, the same in breadth, and about two feet long, bricked up and covered with a stove top, with a griddle for cooking. The draft and the chimney are both on the outer and opposite sides of the tent. This is a most admirable arrangement, drying and warming the whole floor of the tent, which is, of course, mother earth. We sleep in a rough bunk, about three feet wide, and have a good straw bed and plenty of quilts to keep us warm.

It is very warm and comfortable. We have had no severe weather as yet. I cannot realize that it is midwinter.

I must tell you how I passed my Christmas. About noon, in company with Lieut. Graham, I set off for the land of "Secesh." In galo

ing down Pennsylvania Avenue, we were brought up standing more than once by the threatening bayonets of the patrol for the streets of Washington. The city is under strict martial law, and the streets are lined with soldiers, stationed as sentinels. They stopped us only to warn us not to ride faster than a trot, and then suffered us to pass on.

* * * * * * * *

Our journey took us through the far-famed settlement of Falls Church. Add two miserable churches to that of Perry, and let that place run fifty years without repair, and you have a picture of the village of Falls Church. * * * * * *

Yesterday everything wore a gala-day appearance. Almost every entrance to an encampment was arched with lofty and beautifully woven evergreens. The camps were laid out in streets, and thickly strewed with fresh spruce shrubs. They were picturesque beyond anything I ever conceived in camp life; and withal the tents were clean and apparently comfortable, and the soldiers cheerful and contented.

" Who wouldn't sell a farm to be a soldier."

CAMP CONGREVE,
Washington, January 20th, 1862.

Since writing you last, we have had one grand, constant, continual rain—what the boys call a "bully rain." Day and night, pat, pat, patter, it has come on our canvas roofs, compelling us to hover close around our tin stoves, and avoiding the treacherous soil without. Drilling and general camp duties have been almost entirely suspended. We have had nothing to do but make ourselves as comfortable as we could under the circumstances, which we have done. Of course the weather has had an unfavorable effect upon the health of our boys. There is considerable sickness in our Battalion, as well as in this entire division. The mumps, measles, colds, fevers, &c., are giving the surgeons plenty of work. The lighter cases are all treated in the camp hospitals, while severe and protracted diseases receive treatment in the general hospitals. But we are all hoping for dry weather, and a resulting improvement in health. Until the roads are hardened, an advance from this point would be wholly impracticable.

Our troops might succeed in storming earthworks, but must succumb to this accumulation of mud. This Battalion has engaged only in the dismounted drill, and has attained considerable proficiency in ordinary tactics and sword exercise. The boys swing their sabres as lustily as Don Quixote flourished his trusty blade when fighting the windmills. In a few weeks we hope to educate man and horse in the use of gun carriages and rocket tubes. * * * *

CAMP CONGREVE,
Washington, February 1st, 1862.

* * * * Life in camp is just now monotonous. We are in what is known as a camp of instruction, as distinguished from camp in campaign. We have been drilled in the school of the cannoneer dismounted, and as we have not yet received our horses or cannon, we have not yet commenced the regular artillery drill. We, however, expect our horses in a few days, when we will find three months hard labor before us at least. Two incidents have taken place since I last wrote you, which were out of the usual course. We were paid off— an occasion of great interest, and one which gave general satisfaction; the other was the receipt of the provisions forwarded by the worthy people of Perry. The soldier I take it, from what I have seen, does not abound in demonstrations of gratitude, and yet feels as keenly, perhaps, as he who is more loud in his expressions. Carefully drawn resolutions of thanks might sound well, and perhaps make the donees appear to advantage, but they could not add to a general feeling of gratefulness entertained and manifested, particularly by those acquainted with the individuals who contributed. All of the articles reached here safely, and the most of them have been disposed of. The jelly and preserves are to be kept for the sick, and a few other articles we yet have on hand.

* * * * There has been some sickness with us, but now our camp is unusually healthy. We have had a great deal of rainy weather, and Washington has been a sea of mud. Fortunately our camp is so situated that it is less damp and wet than any of the camps around us. I see no prospect of an immediate advance. The bad roads, I think, would alone prevent it. Even about the city the roads are almost impassable. * * * *

CAMP CONGREVE, WASHINGTON, D. C.,
March 10th, 1862.

A long train of ambulances yesterday, crossing the Long Bridge, put me in a gloomy and reflective mood, from which I have not yet recovered. Fancy cannot picture, nor imagination conceive, the horrors of war. That the best faculties of man, his utmost ingenuity, should be taxed to produce weapons for the destruction of his fellow-creatures, is, when we come really to think of it, appalling. Some fiend, one would think, invented the shell, some of them eleven inches in diameter, filled with nails, pieces of iron and balls, which bursting, sometimes kill fifteen or twenty men. The heartlessness of war, and particularly this war, it seems to me, is touched off to the life in the following lines. They are from *Vanity Fair*, and may not have met the eye of all of your readers:

The Song of the Ambulance.

Let the broad columns of men advance!
We follow behind with the ambulance.

They lead us many a weary dance,
But they cannot weary the ambulance.

We rattle over the flinty stones,
And crush and shatter the shrinking bones.

Here we ride over a Christian skull—
No matter, the ambulance is full.

Behold! a youthful warrior is dead,
But the wheel glides over his fair young head.

See smoke and fire! hear cannon's roar!
Till the bursting ears can hear no more.

Till the eyes see only a sky-blue frame,
And a lurid picture of smoke and flame.

And the air grows dense with a thousand sighs.
And shrieks defiance in shrill death-cries.

And blood lies black in horrible streams,
And we think we are dreaming fearful dreams.

But our wheels are strong, our axles sound,
And over the sea we merrily bound.

What do we care for the bursting shell?
We know its music, and love it well.

What do we care for sighs and groans?
For mangled bodies and shattered bones?

We laugh at danger and scorn mischance,
We who drive the ambulance.

Through rattling bullets and clashing steel,
We steadily guide the leaping wheel.

Writhing in agony they lie,
Cursing the ambulance, praying to die.

While some in dreary, death-like trance,
Bleed life away in the ambulance.

Hurrah! hurrah! up bands, and play!
We're leading a glorious life to-day.

For war is play, and life a chance,
And 'tis merry to drive the ambulance.

HEAD-QUARTERS BATTERY B, On Board
Schooner New Jersey, Chesapeake
Bay, May 2, 1862.

Wednesday, April 30th, after having been under weigh for some time, we were obliged to stop because of the fog. It, however finally cleared, and we continued on until 9 P. M. During the day we passed in full view of the old rebel batteries on Pig Point, Stony Point, Acquia Creek, &c. The scenery along the banks of the Potomac is very pleasing, and as you approach the mouth of the river, it deserves more the name of bay than river. One thing that struck me as peculiar, was the fact that there was not a single village to be seen from Alexandria to the mouth of the river. At nine o'clock that evening we were in sight of the lighthouse on Point Lookout, the extreme point of land between the Potomac and Chesapeake Bay. The wind blowing fresh, however, we were towed into St. Mary's Bay, and the next morning, as it still blew from the east, the intrepid navigators thought it not prudent to start out.

Soon after breakfast, Lieutenant Cady and myself, armed with revolvers, our skipper with a short gun, and two sailors with oyster tongs, set forth on a voyage of discovery. We sailed up the bay about two miles, firing at sundry ducks, gulls, loons, &c., without damage to the birds, when we came upon excellent oyster ground. We fell to, and in a short time had five or six bushels of fat bivalves on board. At a short distance off stood a large, comfortable looking mansion, so I went ashore, and strolling up through a well-planted, well-cultivated garden, was met by the proprietor, a well-to-do man (both physically and financially too, I imagine), and very coolly invited me in. I declined, when the old gentleman, waxing more cordial, insisted upon my going into the house. I complied, and a spacious old mansion it was, too, a place where dwelt genuine comfort and good cheer. From the gentleman's conversation I very soon saw that he was a Secessionist, though he tried to talk Union. His name was Col. Cood, and he was very loquacious, and quite at home on the slavery question. He had thirty slaves himself, and not one of them could be induced to leave him. I refrained from telling him what was a fact, viz., that an hour or two before, two of his best "boys" had been pleading with us to take them with us. For fear it might go hard with the "boys," I declined the Colonel's pressing invitation to stay to dinner, but upon leaving he gave me a half bushel of excellent asparagus, and nearly as much lettuce.

Just at evening we rowed to Georgis Island, in quest of sweet potatoes. Several of the boys went with us this time, and while Capt. Long, the Lieutenant and I went to the houses for our vegetables, they went in another direction. What they did or where they went I know not, but one of them carried a revolver, and on our way back I thought I perceived the smell of fresh meat, and this morning we had some very nice veal.

IN TRANSPORT OFF NEW BERNE,
May 12th, 1862.

On the 25th of April we received orders, which were unmistakably earnest, to get our batteries in a state of readiness to proceed to North Carolina, to join General Burnside.

Saturday, 26th, we embarked in five schooners, and early the next morning dropped down to Alexandria, where we lay until Monday, 28th, when we were taken in tow by a steamer. Besides the Battalion the Third New York (Van Alen) Cavalry left Alexandria for New Berne.

We reached Fortress Monroe Saturday, May 3d, having been detained by foggy weather at the mouth of the St. Mary's river about forty-eight hours. We remained at the Fortress until Tuesday morning following, which enabled many of us to go ashore. We found the Old Point what Washington had been all winter, but what it has now ceased to be, the centre of active military movements.

The east wind which had kept us at the Fortress, on Tuesday, 6th, gave way to a nor'wester, when we put to sea. After two days' sail we reached Hatteras Inlet, the only entrance from the ocean through that vast shoal of sand which stretches from Cape Henry southward.

The weather was pleasant and the sea smooth, so that we had but little sea-sickness. The few who were affected, however, had it terribly, which gave occasion to a remark from one of the afflicted, that next to unrequited affection, there is nothing that unmans one like sea-sickness.

The coast here is famous as being the most stormy on the seaboard. For three weeks or more, Burnside, with his fleet, previous to the taking of New Berne, were blown about, and by great good luck escaped a total destruction. The entrance into Pamlico Sound at the Inlet is very narrow and very shallow, and vessels can only get through with the wind in particular quarters. Fortunately our fleet, with the cavalry and transports with stores, about thirty sail, came through safely. After a stay of twenty-four hours at the Inlet, we set sail for this place, where we arrived last night, having been delayed by head winds.

It is just two weeks since we left Alexandria. Our horses have fared pretty hard, and show the effects of confinement. We have been tossed about till we are tired, and rejoice at the prospect of speedily disembarking. We were hailed by friendly voices on our arrival here, and discovered that we had been outsailed by the cavalry, and Zeb. Robinson and Mort. Post welcomed us to the land of Secession.

The steamer, you know, left us at Fortress Monroe. The channel is narrow; the Neuse River is shallow; vessels drawing more than seven feet water not being able to get up here.

The rebels are entrenched, we learn, within ten miles, about 10,000 strong. There are but two batteries, it is said, in Burnside's command, and it is not likely we shall remain here many days, as the policy of Burnside is certainly offensive. The Twenty-fifth Massachusetts, from Worcester, is located here, and the Nineteenth New York (Cayuga) is in the vicinity. Tom Post, of Perry, is in the last named.

One of our men, John Quinn, whose family resides at Portageville, was terribly wounded in an affray on the eve of our leaving Washington. Little hope was entertained of his recovery, and he was left behind.

At the time of the arrival of the Battery in New Berne, General Burnside was in command of that department. It will be remembered that it was General Burnside who commanded the expedition which captured New Berne and Roanoke Island, and opened that part of North Carolina to our vessels and troops. At this particular time his force was small, and there was no reason for supposing that there was to be no immediate advance on the part of the Federal army.

The troops which had been sent to him, including the battalion of artillery, were evidently intended to be used for the defence of this stronghold. It was a point gained, and for the present must be retained.

Battery B consisted then of four three-inch rifled pieces and just men and horses enough to work them. It was encamped out of the city, across the Trent River, on the sand plains (a particularly unpleasant location on a windy day). From various causes, each of the two batteries in the battalion had diminished in numbers. As a whole, they would have no more than properly manned one six-gun battery. At this time Captain Lee was sick and unable to attend to his official duties. There was a great deal of discontent among the members of the battalion. Their commander inefficient and given to intoxication. There was little discipline in the battalion, and it finally culminated in this manner: On the 4th of June a letter of resignation was written and signed by all the non-commissioned officers of Battery B, and sent, through Capt. Lee, to the major. A copy of the letter was also sent to General Reno, who commanded the division in which they had been placed.

The matter of the correspondence was a complaint of want of discipline and of acts of injustice, deception, unredeemed promises and various minor difficulties.

About this same time, Captain Ransom, of Company A, who by rank was in command of the battalion (Major Lion having been, without the knowledge of the men, dismissed), endeavored to force the members of Company B to consolidate with Company A.

On the 27th of June, forming the company in line, he commanded them to answer to the roll call as members of Company A, or to step one side and be taken to the guard house. It must have surprised him somewhat when every member, as his name was called, stepped aside and answered "guard house." This attempt at consolidation failed.

General Reno came to them and addressed them. He heard their complaints, and then informed them that Major Lion had been dismissed from the service for incompetency, and that as fast as he found the remaining officers incompetent, they would be dismissed. He reprimanded them for their insubordination, but gave them to understand that they should have their rights.

A few days after, Company B was placed in the Third Division and Company A in the Second Division, as independent four-gun batteries—named respectively Captain Lee's Battery and Captain Ransom's Battery. On the 3d of July, 1862, Captain Lee's Battery was ordered to Newport Barracks.

Newport Barracks was an outpost, and their duties began to be like those of a soldier. •

For the present we leave, then, this nucleus of the Twenty-fourth New York Battery, and return to Perry, to speak of those who were enlisting to go to Newport Barracks and fill up their ranks.

CHAPTER IV.

THE TWENTY-FOURTH NEW YORK INDEPENDENT BATTERY.

The Union army was meeting with defeat and loss of men. The President made a call for three hundred thousand more. The smothered fire of patriotism that was burning in the hearts of the young men in Perry burst forth, and fathers' commands, mothers' warnings, nor sweethearts' pleadings and caresses could avail aught in trying to subdue the flame. 'Twas contagious, and spread with such uncontrollable rapidity that in a short time about sixty of the bravest and the best young men in that town and vicinity had come forward and enlisted in the cause. Few of the residents of that quiet little place, September 10th, 1862, can forget the morning of that date. To so many homes had the night been long and of little rest; in so many were saddened, heavy spirits and grief-stricken hearts.

The writer has only a confused recollection of tearful faces, of heart-wrung sobs, of sad adieus and fervent "God bless you's."

Full of the ambition and pride of youth—full of patriotic fervor, and eager for the strife—believing we could help to redeem what others had lost—we did not stop to think or realize how true might be our parents' predictions, or the fears and presentiments of our friends. What a blessing to man is ignorance of the future!

On the 22d of August, Mr. George S. Hastings received authority to raise recruits to join the organization called Captain Lee's Battery, then stationed at Newport Barracks. In one week fifty men had volunteered; another week increased the number to sixty.

The citizens of the town where they enlisted, encouraged them with kind acts and kind words. There was a great deal of enthusiasm exhibited throughout the vicinity. Generous bounties were offered and paid. To many of the volunteers this was useful in the final settlement of their pecuniary matters. To the families of others it left a competence for a short time. To all it was acceptable; but to few, if any, was money a motive power to volunteering.

These men, with but few exceptions, were young, and the galaxy of the towns in which they lived. Their enlistment seemed a spontaneous outburst of the single thought that had dwelt in many minds with equal power, "Young men for war, old men for council."

Resolution, courage and determination were stamped in the faces of all. Like the clans of the feudal times of old, they meant to show that the flower and the pride of the country would win the crown of victory or death, and like those stories of old, the long, long days passed slowly by, the weary home watchers waited, hoped, and feared, till finally a remnant few returned in a pitiable plight, to bear the sad tidings of defeat, of suffering, and of death.

It is no more than justice to a few who were unable to

pass the surgeon's critical examination, and who, notwithstanding, would have made capital soldiers, to say that they, too, may be included in that honored list. Their intent was positive, and it was with extreme reluctance that they submitted to his decision. One of the parties wept over his failure to pass, grieved and chagrined at being deprived from accompanying his fellows. Their names were: Norman Macomber, E. H. Andrus, F. A. Calkins, Ezra Higgins and Seymour Sherman.

The Wyoming *Times* furnishes the following interesting accounts:

> The unexpected success of Mr. Hastings in obtaining recruits, induced him to name Saturday as the time for going to Buffalo, and being mustered in so as to secure the State bounty, which at that time it was supposed would cease on the first of September. On Friday his recruiting rendezvous was thronged with persons who wished to enlist, and at times two or three persons were making out the necessary papers.
>
> There was a meeting in Smith's Hall on that night, but most of the young men who came to hear the speeches enlisted before they reached the hall, for the enthusiasm in the recruiting office was much greater and more hearty and unanimous than at the meeting. Before closing the office that night, the list of Mr. Hastings showed that sixty-six recruits had been obtained for Captain Lee's Battery, the majority of them belonging to this town, and all recruited in a little more than two weeks. This result was no doubt owing to the entire unanimity of action which prevailed, and the unceasing energy and zeal displayed by all interested.
>
> MUSTERED INTO SERVICE.—The recruits which have been obtained for Captain Lee's Battery, some sixty six in number, by George S. Hastings, Esq., started early Saturday morning for Buffalo, to be mustered into service. Our townsmen volunteered to take them to Castile, and, altogether, there was quite a large procession of them.

They arrived in Buffalo at ten o'clock, and forming into line at the depot, marched directly to the examining surgeon's office, opposite the post office. The examination was finished about one o'clock, and only eight out of the whole number were rejected. The surgeon was quick and skillful, not a defect of any kind escaping his notice, and so strict as to reject one man that had once been accepted at Portage. He remarked several times it was one of the finest companies he had examined, and in their entire march through the city they were complimented for their fine looks, their seeming intelligence and gentle manly bearing. From the examining office they marched to the Franklin House and took dinner, As soon as this was finished they were called to the mustering office, near the canal, over the Marine Bank, and papers being all ready, were sworn into the service by Lieutenant Sturgeon, of the regular army.

A SPLENDID LOT OF MEN.—A squad of sixty-two men from Perry, Wyoming County, arrived in Buffalo this morning, under command of Lieutenant Hastings. They were recruited for Battery B, raised in Wyoming County, now at Beaufort, N. C. " The boys sent us word they'd like a little help," said one of the men to us soon after their arrival, " and we thought we'd go down and help 'em."

The men came here to be examined and mustered into service, when they will return to Perry. They will probably leave for the seat of war in two weeks.

The following is the list as it stands on the muster roll:

PERRY.

Mason C. Smith	Rufus Brayton
Chas. H. Dolbeer	B. F. Bachelder
Phares Shirley	Edward Welch
J. W. Merrill	Jonas E. Galusha
Chas. H. Homan	Abram Lent
Wm. S. Camp	Jas. Calkins
Oliver Williams	John Filbin
Geo. S. Atwood	Le G. D. Rood
Chas. W. Fitch	P. J. Stafford
Philander Pratt	John McCrink
Porter D. Rawson	Henry Tilton, Jr.
Paul Calteaux	G. W. Keeney
R. H. Barnes	A. W. Comstock
Albert Griffith	B. H. Hollister
Thos. Fitzgerald.	

MOSCOW.

Jerry McClair
L. H. Lapham
R. J. Newton
Horace Lapham
John A. Brooks
And. T. Ferguson
Wm. Carnahan

Chas. A. Marean
Edwin L. Boies
Jacob H. Weller
Jas. W. Perkins
Henry V. Clute
Emmett Wood
Geo. W. Piper

James Button

MT. MORRIS.

Charles McCrary
W. A. McCrary
L. H. Shank
Wm. Blood

J. H. Armstrong
W. M. Hoyt
Hiram Loomis
Edwin Eastwood

CASTILE.

Chas. Hathaway
Hartwell Bartlett

Wm. F. Hosford

E. T. M. Hurlburt, Warsaw   A. L. Culver, Gainesville
John Baker, Covington

TOWN BOUNTY FUND.—The following is a correct list of the contributors to this fund for the town of Perry. The subscribers are requested to pay immediately to G. C. Chapin or C. W. Hendee, at Smith's Bank, who will pass it to the credit of G. C. Chapin, treasurer. It is designed to pay this bounty to volunteers to-morrow or Monday.

Andrus, C. P.	$50	Buell, Richard	$50
Andrus, Martin	50	Butler, N. & W.	65
Alverson, Richard E.	50	Butler, John M.	50
Andrus, Samuel	50	Bradley, John R.	50
Atkins, Martin	25	Brigham, L. M.	10
Armstrong, Sanford	50	Buckland, J. P.	20
Bullard, F. O.	50	Bradley, Edward	20
Bailey, John H.	50	Butler, Aaron	35
Barber, S. R.	50	Benedict, William	50
Benedict, C. J.	50	Bull, Charles	50
Brigham, R. W.	50	Bradley, G. F.	50
Bills, Edmund C.	50	Birdsall, N. D.	10
Brigham, H. A.	50	Benedict, Samuel S.	5
Buell, Columbus	50	Bradley Mrs. A.	5

TWENTY-FOURTH NEW YORK BATTERY.

Name	$	Name	$
Benedict, C. G.	$50	Loomis, H. C.	$25
Brown, Joseph W.	50	Lapham, Alvah	50
Currier, N. P.	50	Lane, Austin	20
Crocker, M. N.	50	Lacey, Henry	5
Chapin, G. C.	50	Leffingwell, Henry	20
Cole, Alexander	50	McEntee, A. S.	50
Corbit, Mrs.	50	McIntire, J. W.	30
Copeland, John	25	Macomber, Allen	50
Chase, E. N.	20	Moffet, R. S.	50
Coleman, G. W.	50	Means, Rev. G. J.	25
Coleman, John	50	Mathews, Edward G.	50
Crichton, Wm.	50	McIntire, Ben. D.	50
Cadwell, Francis	10	McCall, Thomas	15
Chappell, Abner	30	McDonald, John	5
Chappell, Lyman	10	Mace, Eli B.	25
Crabb, J.	10	Miner, Ichabod	20
Cornell, Harrison	25	Martin, Esther	25
Dolbeer, Wm. K.	50	Martin, John J.	25
Daily, J. M.	50	Nichols, G.	10
Davis, M. G.	50	Noyes, Edward	5
Davis, Thomas	50	Olin, John	75
Errickson, D. W.	50	Olin, William	50
Ferguson, Jerome	10	Olin, G. B.	75
Grieve, George W.	50	Olin, Milo	60
Griswold, J. R.	50	Olin, Paris	50
Gay, Norris	20	Olin, Philip	15
Higgins, R. D.	50	Palmer, Tyler	50
Hendee, C. W.	50	Page, H. N.	50
Handley, Jonathan	50	Paterson, T. J.	50
Hatch, Samuel	50	Philips, L. M.	50
Hitchcock, J. B.	50	Pratt, Joel T.	10
Hosford, Nelson	50	Pratt, D. B.	10
Hart, Wm.	10	Palmer, Wm.	50
Hawley, Wm. H.	22	Pratt, R. B.	10
Howard, S. M.	25	Pratt, Jabesh	10
Hosford, Almer B.	25	Philips, W. A.	15
Higgins, M. D.	10	Page, Benj. F.	5
Justin, Proper	50	Philips, Caleb	30
Jeffers, Eugene	30	Pennock, Alexander	5
Judd, B. H.	15	Partridge, Levi B.	20
Keeney, G. L.	50	Reed, Daniel F.	50
Karriger, James	50	Rudgers, John	5
Kingsley, Elias	15	Rouse, Mrs. S.	2
Kingsley, George	10	Sweet, German	50
Kniffin, J. B.	05	Sharpsteen, Mortimer	50
Kniffin, Thorn	50	Scranton, H. M.	50
Lillibridge, H. H.	50	Stedman, R. H.	50

Sherman, J. B.	$50	Taylor, D. R.	$25
Seymour, N.	50	Velzey, M. N.	50
Sheldon, Andrew	50	Williams, M. C.	50
Sheldon, G. K.	50	White, J. H.	50
Stowell, David	25	Wyckoff, J. W.	50
Sweet, Rufus	10	Wygant, E. H.	50
Strong, Albert	50	Wright, A. M.	50
Strong, H. L.	50	Wallace, J. D.	25
Skinner, B. C.	50	Williamson, James	10
Stowell, Hall	50	Watrous, Charles	50
Shaw, D. M.	20	Witter, F. S.	50
Stamp, G. & I.	25	White, G. C.	50
Saxton, Uriah	25	Westlake, J. S.	10
Spear, Wm. B.	15	Winter, John	25
Sleeper, Jonathan	25	Westlake, Albert	10
Sharpsteen, Samuel	50	Wellman, Harvey	25
Tuttle, R. T.	50	Wellman, Nelson	20
Taylor, Steven	50	Wygant, J. L.	25
Tewksbury, S. W.	50	Watkinson, Wm.	10
Toan, Austin	50	Wright, G. H.	20
Taylor & Nobles	50		

OFF FOR THE WAR.—The men recruited by Geo. S. Hastings for Company B, Rocket Battalion, Captain J. E. Lee, took their final departure for the seat of war, Wednesday morning. The citizens turned out early in the morning, to bid a last "good bye" to the boys; and at about six o clock they were on the way to Castile Station, where they were detained nearly three quarters of an hour—the train being behind time. Finally the cars came, and the boys, with quite a large delegation of our citizens, took passage for Buffalo, reaching there a little after eleven o'clock. Forming into line at the depot (preceded by Aplin's Band, who kindly volunteered for the occasion), they marched to the Arcade, where a number of new recruits were mustered in, and then proceeded to Roth's Hall, on Batavia Street, and took dinner. Expecting to leave for Albany the same night, at four o'clock they returned to the mustering office, to receive the Government bounty; but the officials were so busy they had to postpone their departure till Thursday, and they went back to their quarters on Batavia Street, where they expected to have a good time until ready to leave the city. The procession attracted much attention, and many flattering remarks were made by citizens all along the route, complimentary both to the men and the band. We noticed that a number of the boys had bouquets, showing that if they had left home, they were still among friends. On Tuesday, Captain Burt, of the Silver Lake House, tendered the hospitalities of his grounds and steamboat to the volunteers, for pic-nic purposes; which seemed to be "hugely" enjoyed by those present, and all left, wishing the Captain a long life and many such happy reunions. In the evening, at Wallace's Hotel, "Andrew's" Cotillion Band claimed the attention of a large number of both sexes who were wont to "trip the light-fantastic toe"—and altogether the soldiers had "a time" which will, no doubt, be long cherished among those "pleasant memories," which revert to the scenes of long agone. May our best wishes that they all return again to their homes and friends, sound and well, speedily be realized.

After a few days' stay at a German hotel in Batavia street, Buffalo, where we were initiated into rations of Dutch bread, Bologna sausage and lager beer, furnished by the United States at thirty cents per diem, we were sent to Albany.

In this city we were quartered at the Asylum Barracks, and underwent another examination. I cannot conceive for what purpose, unless it was to put the fees into the pockets of the post-surgeon. However, his remark was the same as that of the surgeon at Buffalo, that "It was the finest squad of men that he had examined"—all passed.

We were delayed but a day or two. Monday, September 15th, we took the steamer Isaac Newton for New York City.

At New York we were quartered at Park Barracks, which were at that time cleanly and not over-crowded. The soup, coffee and meat were all good. The Croton water was sweet, and the fruit stalls surrounding us were loaded down with the finest varieties of peaches. We enjoyed our short stay here greatly. We received our uniforms, and 'were permitted to roam about the city during the day and visit various places of amusement in the evening.

To show the spirit and feeling prevailing at that time, we quote from a letter written by William F. Hosford to a friend.

<blockquote>
Our boys all appear to have formed good impressions of this city and their quarters. All have throughout expressed their determination to remain till their services are no longer needed by the Government, and if any are sorry they enlisted, they are wise enough to say nothing about it. For myself I am glad that I enlisted, not only as an act of duty and patriotism, but when I did and where I did. My attachment to the Company and our branch of the service increases daily.
</blockquote>

On the 19th of September we embarked on the steamer "Oriole," destined for New Berne. A storm threatened us while near Fortress Monroe, and we ran into that harbor remaining for two days. A grateful retreat, too, it was. The rough sea had given a general experience of sea-sickness, and a quiet harbor restored smiling faces and merry tongues. The demand for fresh food was too great for the immediate supply—and the rush for the fresh bread and molasses that was issued to us as extra rations from the Fortress was overpowering. However, as we weighed anchor next day and again steamed out toward the sea, the demand decreased, and it was not long before a double row of gaping mouths were giving bread and molasses to the fishes.

A few days brought us to Cape Hatteras, and passing safely over the sand-bar, we found ourselves steaming down the quiet waters of the river Neuse. All were interested in viewing a country new to them, and in listening to the explanations and stories of the approach and attack upon New Berne by General Burnside, his base being the fleet of transports and gun boats which were stationed in this river.

There was a general brushing up of new uniforms, appearances on deck of faces that we had not seen for days, and a return of jest and laughter. Our steamer was fastened to the wharf, and after giving hearty cheers for our kind and gentlemanly master, Captain B. F. Holmes, we disembarked. Captain Lee, expecting our arrival, had had railroad transportation to Newport Barracks provided for us, and we were soon in the cars and started off.

Arriving at Newport Barracks we were warmly greeted by our old friends, who were glad to have our assistance and companionship.

In the course of a week or so, Lieut. George Hastings, who had remained in the North to complete some business arrangements, and add a few recruits to the number already obtained, arrived with half a dozen recruits, accompanied by Clark and Lieut. Fred. Hastings, who had also been home on recruiting service.

As soon as these last recruits had arrived, there was a new interest aroused as to the appointment of non-commissioned officers.

There were some conflicting interests between the "Old Boys" and the "New Boys," in the re-organization and these appointments, but time destroyed these phrases of distinction, and unanimity of feeling and interest, or at least as much of it as could be expected in any company, prevailed.

On the 19th of October orders were received, designating the company as the "TWENTY-FOURTH INDEPENDENT BATTERY OF LIGHT ARTILLERY, NEW YORK STATE VOLUNTEERS."

On the 4th of November, the appointment of non-commissioned officers, and the assignment of privates to their particular posts, were made. The following is the roster as read to us at our parade:

Captain.—J. E. Lee.
First Lieutenants.—L. A. Cady, Geo. S. Hastings.
Second Lieutenants.—Geo. W. Graham, Fred. Hastings.
Orderly Sergeant.—C. H. Dolbeer.
Artificers.—A. Richards, A. Griffith, P. D. Rawson, P. Calteaux, M. Grant.
Buglers.—H. C. Burd, L. Newcomb.
Guidon.—Wm. Alburty.
Wagoner.—J. Chapman.

First Detachment.

Sergeant.—R. C. Ainsworth.
Gunner.—F. D. Otis.
Caisson Corporal.—Wm. A. McCrary.

Wm. Ainsworth,	A. McDonald,	J. Flynn,
J. E. Galusha,	M. C. Smith,	T. Fitzgerald,
S. Rowell,	R. Brayton,	George Miller,
M. Ansbacher,	W. P. Nichols,	J. Sunfield,
J. H. Weller,	A. W. Comstock,	H. Bartlett,
	J. W. Perkins,	J. G. Miner.

Second Detachment.

Sergeant.—J. W. Merrill.
Gunner.—E. T. M. Hurlburt.
Caisson Corporal.—G. G. Wright.

R. Bullock,	E. Eastwood,	J. McVey,
R. J. Newton,	W. F. Horford,	C. McCrary,
D. Munroe,	B. F. Bachelder,	C. H. Homan,
E. H. Hunter,	C. W. Fitch,	J. Baker,
O. G. Parmlee,	Le G. D. Rood,	H. Chadbourne,
	A. L. Culver.	

Third Detachment.

Sergeant.—C. R. Griffith.
Gunner.—A. T. Ferguson.

W. M. Hoyt,	J. Button,	L. H. Lapham,
H. Lapham,	E. Boies,	G. B. Johnson.
T. Rich,	W. E. Chapin,	H. Cook,
Z. Allen,	J. H. Armstrong,	M. R. Mosier,
G. W. Stevens,	H. V. Clute.	W. Gould,
	S. King,	T. F. Shockensey.

Fourth Detachment.

Sergeant.—Wm. S. Camp.
Gunner.—Geo. Birdsall.

J. Woolsey,	P. Pratt,	F. M. Alburty,
G. S. Atwood,	B. H. Hollister,	O. Williams,
W. A. Whitney,	M. Andrews,	E. Wood,
F. Leonard,	R. H. Barnes,	J. Calkins,
S. Root,	J. Crooks,	J. McCrink,
	G. A. Holman.	

Fifth Detachment.

Sergeant.—H. P. Lloyd.
Gunner.—B. F. Corbin.
Caisson Corporal.—P. Shirley.

S. Nichols	Wm. N. Page,	H. Tilton,
R. Turner,	S. D. Canfield,	W. W. Crooker,
G. W. Kellogg,	M. Crosby,	C. Marrin,
P. Marrin,	A. Lee,	P. J. Stafford,
J. H. Armstrong,	Geo. W. Piper,	H. S. Whitney.

Sixth Detachment.

Sergeant.—C. A. Clark.
Gunner.—S. Stoddard.
Caisson Corporal.—C. T. Phelan.

W. Blood,	J. McClair,	A. Adams,
C. Hathaway,	O. S. McCrary,	E. H. Wardwell,
G. Barker,	T. McGuire,	Ira Billingham,
J. T. Ferrin,	Geo. W. Keeney,	E. Richards,
J. Cowen,	W. Carnahan,	E. Welch,
	J. Filbin,	J. Russell.

CHAPTER V.

NEWPORT BARRACKS.

After the names of the new recruits had been added to the muster roll, it contained the names of five officers and 126 men.

We had but four pieces, a few horses, and a scanty supply of small arms and equipments. A requisition had been forwarded to the proper authorities, however, and we felt encouraged to hope that we should, before long, be a formidable organization. Drill was made imperative and constant. There was an eager desire on the part of the recruits to learn, and a willingness on the part of the veterans to teach. They had their laughs and their thrusts at each other, but no serious dispute ever arose between them. The experience at Newport Barracks, as a whole, was a pleasant one. There was sufficient exercise in our duties, our food was good, and there was excitement by being on an outpost. Rumors of the approach of the enemy, or of an advance on some rebel grounds, kept us on the alert.

Good barracks and tents were furnished us, and nearly every building had its mess, by means of which a sort of house-keeping arrangement was undertaken—and certainly no soldier could ask for cleaner quarters or better meals than we then enjoyed. There was but little sickness, and the extremely well carried on hospital at Newport, soon put any sick ones on their feet again.

On Wednesday, November 5th, the third and sixth detachments of the Battery were ordered out on a scouting expedition. In addition to these the cavalcade consisted of a company of the Van Alen cavalry, one company of the Ninth New Jersey, and a battalion made from a couple of Massachusetts regiments. Captain Lee was in command of the expedition. At half-past four in the morning the command, "Forward" was given, and on went shuffling feet, clattering hoofs and rumbling wheels.

A short distance from camp one of the caissons, in passing through a narrow space, run one of the wheels upon a stump, completely overturning the caisson, setting the cannoneers flying in every direction, breaking both poles, the reach, stock and trail, and loosening the ammunition chests from their fastenings. It could be of no further use until it was repaired. Therefore the drivers returned to camp with it, while the cannoneers found seats on the gun chests, and went on with their piece.

Nothing further occurred to relieve the dullness of the tramp through the woods, save an occasional return of some cavalrymen, who would bring in some "guerrilla" or "union man," and perhaps an old musket or two. Some bee hives in the yards of the houses they passed by looked rather tempting, and not a great time passed before broken hives were supplying buzzing bees and hungry boys with their delicious contents. But the boys found that the bees stuck by much longer than the honey, and after they had halted and encamped for the night, having had some thirty miles march, a host of these revengeful little fellows were constantly giving a thrust into the face or body of some one of the despoilers of their homes.

That night it began to rain and continued to pour down steadily until the following night. The infantry could not use their muskets on that account, and as appearances indicated continued rain, and it was growing quite cold, it was though best to return.

They reached camp next day thoroughly drenched with the rain, with twelve prisoners, some horses and other contraband articles. This was their first scouting experience. It was enjoyed by all who participated in it.

On the following Wednesday, at about three o'clock in the morning, an extra train came thundering down to the station, and our Captain, who had gone to New Berne the previous day, skipping off the train, called for the "Corporal of the Guard," and told him to "call out the men to hear orders." The unusual commotion routed us from the tents, and in a wonderfully short time we formed the line, and were informed that we were to proceed immediately to New Berne.

From one of the men attached to the train, we learned that the pickets within a few miles of New Berne had been driven in, some of them killed, and an attack on the city was anticipated.

In coming down after the troops, the engineer had taken the precaution to have a hand car run ahead of the engine, fearing that the rebels might have torn up some portion of the track.

As we stood by the camp fire awaiting the arrival of some of the Ninth New Jersey, who had been sent for at

"Bogue Sound," stevedore Gilfillin, who had charge of the "contrabands" that were in the hand car, amused us with stories of their odd actions and expressions, as they moved along through the darkness over the track. Either 'Boneyparte' or 'Washinton' or 'Bolivar' or 'Franklin' were constantly seeing a rebel or a troop of them, "way long on de track." Nevertheless, they assured themselves as well as the stevedore, that they "weren't afraid," and argued with each other as to the best method of carrying on a fight. Suddenly, Gilfillin cried out, "There they are!" "Oh! Lord, Boss, whar?" they ejaculated, and with "*pallor on their cheeks, and their hair standing*" they dropped the wheel handle and made preparations to leap.

It was some time before he could convince them that he "had made a mistake." And he concluded that if he thought of getting to Newport, he should try no more such experiments.

The pieces, caissons, horses and men were on the cars by six o'clock, and off they started with the colors flying, and rousing cheers.

The next day they returned and reported that the whole affair was more "scare" than "hurt." This was their second "call out." And though they did not participate in any battle, yet they fully expected to meet the enemy. It was an excellent opportunity to show their "metal," and there was no lack of it.

About this time Mrs. Lieutenant George Hastings arrived on our favorite steamer "Oriole." So rarely was a female face seen, that she was received with great admiration and a sincere welcome.

In the latter part of November, some of the boys discovered a steam sawmill located within a mile or two of the camp. Our practical Lieutenant Cady proposed to turn the discovery to advantage, and promptly made an examination of the condition of the mill, and reported it soon after in working order. Our drill was thereafter alternated with labor in the woods. Each section of the three went out into the woods every third day to fall trees and float them down to the mill, where Cady, Rawson, Richards, Pratt, Albert Griffith, Rich and Woolsey soon converted them into desirable lumber. The writer well recollects passing an eventful evening, while assisting to float a board raft down to the railroad bridge, from where the lumber was to be shipped by railroad. As we floated down the river, the bright pitch fire upon the raft, casting its lurid light into the dark shadows of the immense forest trees that leaned over the shore of the river—the dark, nearly invisible forms of our comrades that sat with the fire interposing between us—the merry song and the laughter over the comic story—the exclamation over the roasted sweet potato that some one had broken open, while the hot steam and the savory odor rose to the open mouth till the waiting palate danced with joy—all these seem a strange, wild picture that haunts my memory yet.

Nor can we forget the "Halt! who comes there?" and the sudden presentation of the bayonet of the Massachusetts sentinel, and our sudden huddling together and approaching to give the necessary salutation and password. The mill turned out 15,000 feet of lumber, which was a great assistance to us the following winter, in building winter quarters in New Berne.

Tuesday evening, December 10th, an extra train brought down orders for us to strike tents, pack up all our personal effects and be aboard the cars by the time the engine returned from Beaufort.

We were ready at the appointed time. And about one o'clock were on the train, started for New Berne, bidding Newport Barracks "good bye" for ever. The Battery had been stationed there five months. We felt quite at home there, and had not that strong desire for an active soldiers' life been so predominant in our minds, I doubt not we should have felt a little twang of regret at leaving.

We make a few selections from the letters of the correspondent of the Wyoming *Times*, which relate more particularly to their social and domestic life:

HAMMOND HOSPITAL, BEAUFORT, N. C.
October 16, 1862.

Day before yesterday I took a sail of about ten miles with a party of twelve others. There was a good stiff breeze, and we ran at the rate of 2.40, nearly, making the ten miles in three quarters of hour. We stopped on the *sound* side of the island, and then took the direct path that led to the beach. The surge rolled high—and as we went for the purpose of gathering shells—this was very favorable for us. We wandered along the beach toward the Cape Lookout light house, and approached near enough to take a good survey of it. It is one of the best light houses on the shore, but for some reason it is not now used. This Island is named from the cape, Lookout Island. We soon turned into one of the paths (they have no roads here, and the country is entirely covered with a low brush that they term woods, with occasionally a path in it that leads to a dwelling), and following it about a mile we came abruptly upon a settlement, three or four houses with their kitchens and cook room, which are always separate from the house, that is, detached from it. The old lady out on the stoop was considerably frightened at first, at so many blue-coats, and refused to entertain us at all, but one of her sons coming to the house, she finally concluded that she would keep half a dozen of us over night and give us tea and breakfast. The salt water breeze and long walk had given us voracious appetites, and as we sat down to milk, eggs, fish, sweet potatoes and Youpon tea, you would have been astonished to have seen it disappear so rapidly before six hospital patients.

After tea the boys sat down to do their smoking, and the old lady, son and daughter, did the talking. They entertained us with stories of the war—particularly the taking of Fort Macon and Beaufort. It seems that our forces took the place just in time to prevent the son, with sixteen others who had been drafted by rebel conscription, being forced to go into the southern army. There is not a rebel on the whole island, and there was great anxiety when the battle was fought, and great rejoicing when we had won the day. The island was under cross fire of the forces, and many shot and shell fell on it, often quickly dispersing the crowds that had collected to see the fight. Anecdotes were numerous, so much so, indeed, that I should really get the "stories mixed" if I should endeavor to tell any of them.

We rested very well that night, considering that they gave us feather beds to sleep on—for I cannot now rest or sleep on anything softer than a mattress. Next morning we were up early and over to the beach before breakfast. When we returned the old lady had prepared us a fine breakfast. As we offered to pay her for our entertainment, she refused any remuneration, saying, "I want to do all I can to help the Union cause, and you are welcome to all the house affords as long as you choose to stay." We had dinner at another house before we came off, when they made the same remark in substance, and refused to take pay.

I noticed that there was a greater variety of the feathered tribe on this island than on the other shore, as all the birds are there represented in one—the mocking bird.

Yesterday I went to Morehead City, which is situated and bears about the same relation to Beaufort that Jersey City does to New York. It is a small place, mostly inhabited by fishermen. The Ninth New Jersey are barracked there. Most of the best houses, as are those in New Berne and Beaufort, are confiscated and used for officers and some times soldiers quarters. Just as we came up to the wharf, the guard had arrested a man, calling himself a "Union citizen," who had come down from Swansboro, some sixty miles up in the country. The officers, upon questioning him, concluded that he had better remain a while at Morehead, and the boys soon confiscated his boat after hearing the decision. Our pickets and other guard, are on the close lookout for such fellows, and they "bag" a good many of them.

By looking on the map you will see that Beaufort is situated on the point of land running out into the Bogue Sound. Upon the northern point of Bogue Island, and nearly opposite Shackleford banks, is situated Fort Macon. The channel is here very narrow and winding, which makes it difficult for any boat to come into harbor and so much the more difficult for one to run by the Fort. Fort Macon is very little like Fortress Monroe, and I could understand by visiting both of them the difference between a Fort and a Fortress much more readily and distinctly than by the dictionary definition. Most of the soldiers barracked there are Regulars. The day I went out they were target shooting, the target being an old vessel, about two and a half miles distant from the gun. They made some very fine shots. I hardly think a "secesh" vessel will ever pass in the day time and not feel the effects of a well directed cannon ball.

J. W. M.

HAMMOND HOSPITAL, BEAUFORT, N. C.,
October 15, 1862.

Friend Frank: Having been in the hospital for the past week, I have had little opportunity to see or hear anything that would be very interesting for your readers. I have had, though, a very good opportunity to study the nature and character of the native colored inhabitants, as many of them are waiters in the building, and there is not a barn or shed six by eight but is crowded with them till their heads hang out of the windows.

Last Thursday they held a prayer meeting in the wash-room. It is a good sized building and was well filled. There was really music in the hymns they sung. They have (both male and female) soft, sweet, musical voices. The air flats a little, and gives a peculiar accent and upward slide at the end of every line. This is a peculiarity I have noticed in all their songs, whether they be negro melodies or church tunes. They commenced their evening service by singing "Rock of Ages." Their minister, a colored brother, then made a short but able prayer ; another hymn was sung, and he began a short extempore sermon. His text he selected from Jeremiah—" What will it profit a man if he gaineth the whole world and lose his own soul." 'My beloved breveren," said he, 'Jeremiah was one of thirteen brothers, and God took to him case he was always good. Well, one day they were all to work in a field, and Jeremiah was so tired that he went under a tree and lay down and went to sleep. While he was asleep, God come sailing down from heaven, changed into a dove, lit in the tree over Jeremiah's head, and woke him up saying the words of this text." After this explanation, he immediately commenced an exhortation that was full of life and energy, as the roused spirit of the congregation began to testify. The spirit was warmed and commenced to move, and they continued the "good time," a long time after I had retired and fallen asleep.

A few evenings after this, I witnessed a gathering of quite a different character. Upon a large piazza in front of one of the houses opposite the hospital, a large number of negroes had collected, and they gave us quite a " select and amusing Ethiopian entertainment" —solos, quartettes, and choruses. Negro melodies entirely new to my ear, peculiar to themselves, with low, undulating, wailing choruses, sung in good time and with much effect. They closed the evening's performance with a general " break-down," and of all the grotesque, gymnastic, elastic movements that I ever witnessed, that capped the climax. They cannot be imitated with any degree of perfection. As Ole Jim made his last evolution, and sat down astride the banister, with a hearty " Yah ! Yah ! Sam you can't do dat last."—" Ah," says I, " George Christy, you are outdone, outshown ; your light is hid for a time until you practice more."

I can't like these colored people. They are slow, dirty and lazy. They are always happy, full of song and play. They have not the least education, nor do they wish to have—it is too much labor to study. They have a little natural wit, and if there is no work to be done, can aptly understand anything you tell them. They are somewhat superstitious, a few of them religious, but they all are that same class of lowbred, nature-led, indolent human beings, and it is difficult for me to see how some of you philanthropic people north will ever (as you say can be done) make anything else of them.

The steamer United States came into Morehead City, Saturday. She came in to let off Gov. Stanley, who has been on a visit to President Lincoln. She was loaded with troops for Port Royal. Three new regiments are expected at New Berne. Present prospects seem to indicate that we shall have some fighting to do this winter. It looks as if an advance would be made on Goldsboro' and Rolla. I cannot see any necessity for as many troops as are already here, and as reinforcements are constantly arriving, I think it a safe conclusion to say that we shall have the pleasure of seeing active service before long.

Captain Lee is gaining now quite rapidly. He will soon return to duty, and you may expect we shall have work to do.

J. W. M.

NEWPORT BARRACKS, N. C.,
October 25, 1862.

Interesting matter is scarce. In fact, unusual experiences, particularly in the battle line, are something we have not yet met with, nor do we expect to do so for some time to come. Our hardest battles are with chills and fevers and other local diseases. We are coming out victorious, as we have now but three men in the hospital (Charles Homan, Perry; W. E. Chapin, Arcade, poisoned; E. T. M. Hurlburt, Warsaw, chills and fever, none of them seriously ill), and those in camp come out to drill with an earnestness and activity that shows an increasing health and strength "Jack Frost" paid his addresses for the first time night before last. It would be preferable if he would just send in his card, and not come upon one so suddenly, giving one no time to make preparations to resist his "stinging" gripe. It was so cold, that long before morning came many of the boys went out to the guard's camp fire and sat around that, so that they might be warm.

Next day our lieutenant made requisition for stoves, and our tents are now a cozy little dwelling place, furnished generally with home-made bunks, writing table, cupboard and a Sibley tent stove. These stoves resemble an old-fashioned engine smoke stack, of inferior proportions, turned upside down, and only faulty in one respect, and that is the limited accommodations for cooking. Every mess has considerable of that to do, the day being spent in drill, cooking and dish washing, allowing a small space of time for reading and writing. Provisions are high, as the following prices will show: Butter, 40c.; cheese, 25c.; apples, 30c. a dozen; milk, 6c. per pint; brown sugar, 20c.; flour, 7c. Sweet potatoes are our substantial food. We can purchase them at from 50c. to 60c. per bushel.

The "Signal Corps" have been making their head-quarters in our camp for a few days. They are making a line of signal staffs from New Berne to Beaufort. This is deemed quite necessary, as there is no telegraph wire upon the railroad. These staffs are placed within five or six miles of each other and communication carried on with flags. Oftentimes the flags are only visible through a spy-glass. On some accounts these are preferable to telegraph wire, since a guard can be stationed at each one of them, and there are no wires to be cut. The whole line has been surveyed, and men will soon be engaged in putting up the poles. There is a good deal of activity on the railroad just now. Extra trains, loaded with provisions and stores, are running every day, seeming to indicate that there will be a change somewhere before long. We expect to receive orders to move in the course of two or three weeks. I should not at all wonder if an active campaign were carried on in North Carolina this winter.

To-day Lieutenant George Hastings arrived with a number of new recruits, having taken a "tug" at Hatteras and come up the Neuse to New Berne. The remainder of his party went on to Beaufort in the steamer.

J. W. M

NEWPORT BARRACKS, N. C.,
November 25, 1862.

Our commissioned officers are kind, gentlemanly and even forbearing to us privates, though for all that they hold us up none the less strictly to military rules and regulations. They are good, moral, *temperate* men, and waste no time in idleness. Not one of them either would, from personal antipathy or dislike, refuse to any one under his command all the liberties allowed them, or in sickness refuse to do anything in his power to relieve or aid them.

Old feuds are fast dying out, and it is very seldom now that you even hear them mentioned.

There is no dissension concerning officers. I really wonder at the unanimity of feeling that prevailed, as the two companies, one having been a year in service and the other raw recruits, came together and divided up the non-commissioned offices. There was wonderfully little dissatisfaction expressed—hardly any—at the order making the division and appointing the officers of the company, as I gave you in my letter of November 11.

We are neither becoming dissipated, lazy or slovenly. "Camp life develops the bad qualities of bad men, but, on the other hand, it is favorable to the highest exhibition of virtue, of gentleness and of heroism." So says one of the popular authorities of the day, and such in my little experience in that kind of life I find to be true. Drunkenness is a thing almost unknown among us. I never have seen but three men since I have been here that were under the influence of liquor. We cannot very well be idle, as we have five hours' drill per day, beside police duty; and just now Lieutenant Cady, with gangs of men selected from the company, has two sawmills in operation; one upright saw with water power, and one circular saw with steam power.

As for slovenness, I am certain any "committee of housekeepers," upon an examination of our tents, would return a report of "well done for boys." To conclude this subject, I would say, that in conversation with all of the boys, I find them contented and well satisfied with their officers and their associates, and I do not think you could persuade one of them to accept a discharge from the service, unless the war had ended and there was a prospect of peace.

As the winter approaches there is no necessity for our anxious mothers and kind fathers to have any anxiety concerning our being comfortable and warm. Our tents are tight and water-proof. Most of us have taken boards from four-and-a-half to five-and-a-half feet in length, pointed them, and driven them about a foot in the ground in a circle around the tent. We then raised the canvas to the top of the boards, already battened and banked up. This makes the tent very warm and roomy. In ours we built a fireplace. And in the evening the blaze of the pitch-pine lightens as cheerful and happy a picture as any would care to look upon. In fact, I am a little ashamed to own that we are so comfortable. It hasn't the smack of hardship, &c., that we expected and rather desired to experience. It hasn't the dash and the wildness about it that younger persons consider necessary to fill out their idea of a soldier's life.

J. W. M.

NEWPORT BARRACKS, N. C.,
November 28, 1862.

Yesterday was Thanksgiving Day. The day previous, our detachment worked hard in the woods all day chopping logs and floating them down the river to the sawmill. On account of a lack of pork and beef in the commissary department, we fared, while at work, upon "hard-tack" and coffee. As a consequence, we had appetites on Thanksgiving Day that a poor dyspeptic might well envy. The following was the dinner programme:

SOUPS.

Fresh Beef.   Chicken.

MEATS.

Fresh Beef, boiled.   Fresh Pork, fried.
Chickens, stewed.

VEGETABLES.

Sweet Potatoes, boiled and fried.   Onions, boiled and *buttered*.
Turnips, boiled.

EXTRAS.

Coffee.	Fresh Bread.	Tea.
Crackers.	Molasses.	Sugar.
Butter.	Salt.	Pepper.

DESSERT.

Pancakes and honey.   Sweet Potato Pie.
Pitch Pine Gum.   Apples.

If you can beat that "bill of fare," and have an appetite corresponding with it, I'll "cave." All I have to say, to explain the reason for such an abundance of chicken and honey is, that the evening previous a half dozen from our detachment passed guard in and out without any countersign, except the presentation of a couple of chickens to the guard as they passed in.

Passes were allowed to be given by the sergeants at their own discretion. After dinner the whole camp was nearly deserted. Toward six o'clock stragglers marched into camp with chickens, pork, beef, honey, &c., at shoulder arms. It was very fortunate, as the quartermaster's stores are just now at a low ebb.

After evening roll-call, the Massachusetts boys invited us all to come up to the building formerly used as a hospital, and have a dance. The hall was decorated with flags, knapsacks, accoutrements, &c., and splendidly lighted with *three long tallow candles*. The dazzle of gilt, of "straps" and "stripes," was indeed a gay sight. The music was furnished by Ferguson's band. Order was called by the tap of the fiddle bow on the back of the fiddle. Two sets were formed; the head of the room taken by the major. Order and decorum were preserved throughout. The "ladies" received the most obsequious and constant attention. Gallantry, not of the coxcomb order, was the order of the evening. All went "merry as a marriage bell," till the bugle sounded the taps about eleven o'clock, when all quietly dispersed.

CHAPTER VI.

KINSTON, GOLDSBORO' AND WHITEHALL.

On the 11th of December, General Foster, then commanding the department of North Carolina, gathered together his available troops and made an advance from New Berne towards Goldsboro'. General G. W. Smith had been placed in command of the Confederate troops in the same department. They were supposed to number about 12,000. The main object of General Foster was to reach Goldsboro' and destroy the railroads centering at that point. This place was then on the main line of communication from Richmond, south. General Foster's force consisted of between 10,000 and 15,000 men, composed of four brigades; the right commanded by General Wessels, the left by Colonel Lee; right centre by Col. Amory, the left centre by Col. Stevenson. In the centre, unattached to brigades, were Captain Ransom's and Capt. Lee's Batteries; one battery of the Third New York Regiment Artillery, and two sections of heavy guns—one of 32-pound howitzers and one of 20-pound Parrot guns.

The first and third detachments of the Twenty-fourth New York Battery were all that participated in the march. Many men from the other detachments filled the vacancies that occurred from sickness, &c.

Thursday, the 11th, the force marched about fifteen miles.

On Friday, the advance was slow, on account of meeting with fallen trees that the rebels had placed in the roads, and with burned bridges which they had fired as they retreated. There was skirmishing throughout the day, but no regular engagement occurred.

On Saturday, the section of the Twenty-fourth New York Battery, with the Forty-sixth Massachusetts Regiment for a support, separated from the main body, and took, per order, the more direct road to Kinston. They were sent to guard some point of cross roads, which they reached at about twelve o'clock that night.

Sunday morning, the Forty-sixth Regiment and the third detachment returned by a short road to the main body, and a company of the Third New York Cavalry was sent over and joined the first detachment of the Battery.

The third detachment was sent in another direction, to guard a bridge at which there had been some skirmishing the day previous.

The first detachment continued its march, and about ten o'clock, as they approached a small creek, they discovered about 1,000 rebel infantry and a detachment of artillery, prepared to dispute the pass. The cavalry dismounted, deployed, and, with their carbines, acted as a support to the Battery. Our boys then opened on them with shell. After about an hour's fighting the enemy retreated, taking with them their killed and wounded. After they were positive that the enemy had gone, the Federals rebuilt the bridge that had been burned, crossed it, and arrived at Kinston at four o'clock. This was their part of the battle of Kinston.

The battle by the main body was fought on another road—the enemy gradually falling back, until they came to the bridge, at which time the Ninth New Jersey charged and took two brass guns. While crossing a bridge, Colonel Gray, of the Ninety-sixth New York, was instantly killed by a musket ball, which struck him in the breast and passed through his heart. The rebels attempted to fire the bridge, but failed. Six guns were taken by a company of our cavalry on the other side of the river. That night the whole army entered Kinston. Through the main street a pile of cotton, reaching nearly a quarter of a mile, was burning. Many other things the rebels had set fire to, and the inhabitants had almost entirely deserted the place. Some of our soldiers went to the extreme in plundering the houses and stores.

Monday, the army recrossed the bridge at Kinston, which was a very long one. After they had all passed over, the bridge was burned. They then pressed on towards Goldsboro', making a march of about fifteen miles that day.

Tuesday, a sharp and brisk fight occurred near Whitehall: Our boys were under fire nearly three hours. Finally the rebels retreated, and our army destroyed two new gunboats which were in process of building.

Wednesday, they advanced to the bridge at Goldsboro'. Here quite a severe action took place. Many were killed and wounded on both sides. The United States troops succeeded in burning the railroad bridge and tearing up about five miles of the track. General Foster learned that the Confederates had concentrated a superior force at Goldsboro', and determined that it was unwise to make any further advance. The next morning, therefore, the line of march faced homewards, and a rapid retreat was made to New Berne.

They reached that place on the 24th, having been about ten days and marched about two hundred miles. The Federals lost 90 killed and 478 wounded. The Confederates lost 71 killed, 268 wounded and 476 prisoners, most of whom were immediately paroled.

The following order was afterwards read to the troops in General Foster's command:

HEAD-QUARTERS EIGHTEENTH ARMY CORPS, }
NEW BERNE, January 15, 1863. }
[General Order No. 18.]

In consideration of, and as a reward for, their brave deeds at Kinston, Whitehall and Goldsboro', the Commanding General directs that the regiments and batteries which accompanied the expedition to Goldsboro' inscribe on their banners those three victories—
Kinston, December 14th, 1862.
Whitehall, " 16th, "
Goldsboro', " 17th, "

The Commanding General hopes that the future fields may be so fought that the record of them may be kept by inscription on the banners of the regiments engaged.

By command of Major General J. G. Foster.
SOUTHARD HOFFMAN,
Asst. Adjt. General.

CHAPTER VII.

1863.

NEW BERNE.

Soon after the return of the troops from the expedition to Goldsboro', the Twenty-fourth New York Battery received six very handsome new six-pounder Napoleon guns, an additional supply of horses, new harness and new equipments throughout. The men, consequently, had plenty of employment in breaking in the new horses and engaging daily in battery drill. Prior to this, detachment and section drill had been all that we had been taught. The rapid and sometimes intricate movements of the light artillery battery in a field require practice as well as coolness and skill in execution. For instance, the command "Left Wheel" is a simple ejaculation. But let us view a battery as they execute it—six guns positioned in line, abreast at intervals of four yards, with six horses attached to each gun. Three yards behind each gun stands a caisson, with six horses attached, a rider to each span of horses. The bugle sounds the command. Immediately the left piece becomes a centre, on which turns the long sweep of horses, pieces and caissons. As the distance from the centre increases, so proportionally must the rapidity of motion increase, to keep up an unbroken line. Imagine now how swiftly must the piece on the extreme right move, to retain its position and its distances. If your mind cannot comprehend it, sit on the extreme end of a whirlagig, and ride once around the circle.

Horses enjoy the excitement, and they learn to know different commands of the bugle. The men, too, become aroused and interested, and the maneuvers of a well-drilled battery are a pleasing and exciting sight to any one. The Battery boys were now beginning to taste a little of the experiences that they had read of and hoped to participate in. They were proud of their organization, and had good reason to be.

The city of New Berne was being strongly fortified. A new parapet was thrown up in nearly a continuous line from river to river, enclosing the entire city. Fort Totten, in the centre of the line of fortifications, was large, and filled with heavy artillery. There were also many smaller forts.

The major portion of the artillery belonging to the command was stationed near Fort Totten. The Twenty-fourth Battery was on the hill at the left of the fort. Here, with the lumber which had been obtained by working the sawmill at Newport Barracks, we put up some substantial stables, cook-house, &c. At the same time provided with plenty of tents, we made very comfortable quarters for ourselves.

On the 17th of January an expedition was sent from New Berne to Trenton. Camp furnishes us with the following account of it:

13

The force consisted of the Forty-third, Forty-fifth and Fifty-first Regiments, Mass. Vols. Infantry ; eight companies of the Third N. Y. Cavalry ; one section of the Twenty-third N. Y. Indpt. Battery ; one section of the Twenty-fourth N. Y. Indpt. Battery, and a small force of engineers. The entire expedition consisting of 1,500 infantry, 600 cavalry, 100 artillery and engineers.

Lieutenant Colonel Emmory, Mass. Vols., commanded.

The object of the expedition was to make a feint on Goldsboro' and Warsaw, and thus detain troops which might be sent to Richmond or Petersburgh, to resist a movement, which was planned by our forces in that vicinity. The expedition moved from New Berne on Saturday, January 17th, 1863, at 6 o'clock, A.M., and camped that night at Pollocksville, a place sixteen miles from New Berne, situated on the Trent River, and then containing about twenty houses. We were obliged to encamp here, because the enemy had obstructed our further progress by felling trees across the road, as they retreated before us.

On the 18th we proceeded to Trenton, where we arrived about two o'clock, P. M. As our cavalry approached the town, they were fired upon by some rebel cavalry, who were endeavoring to cut away a mill dam, and thus swell the stream, and detain us. But a few shots from the cavalry howitzer caused them to beat a hasty retreat, and as we entered, they left the town. We encamped here that night, and the next morning burned two bridges over the Trent River, the jail, a grist and saw mill. And after we had crossed the stream, tore open the mill dam, and returned to Pollocksville, where we arrived about three o'clock, P. M.

It was while crossing this stream, which was swollen to the horse's belly, that Benjamin Hollister, who was driving the middle team of the gun, happened to sneeze out his upper teeth—poor Ben.—a sicker looking mortal never existed. He proposed to have the troops stop, and look for his teeth ; but in a glance saw that they were gone from him for ever—and he, three days from camp, doomed to gum it on hard tack, or perish. Self-preservation, that great first law, was adopted by Ben., and accordingly he stopped at many houses on the road to get meal to make soft bread and mush of, and in this way worked through until we got into camp again. At Pollocksville we encamped again for the night, and having on the first night's encampment burnt all the fence rails and cleared things generally, all that now remained to complete the ruin, was done during the night by the troops engaging in that innocent amusement of burning buildings; and on the next morning five or six buildings were all that remained of Pollocksville.

On the 20th, we marched to Young's Cross road, on the White Oak River, ten miles from Pollocksville, arriving there about noon. The enemy had destroyed the bridge, but our engineers soon constructed a suitable one, and our cavalry crossed and started for Jacksonville, leaving the infantry and artillery. On their way to Jacksonville (which is on the New River, twenty miles from Young's Cross Road,) they met and engaged the enemy in a running fight, for nine miles, losing two men killed and taking but one rebel prisoner. The rebs burned the bridge, 150 feet long, over the river at Jacksonville, to prevent our troops crossing. We encamped here for the night, and it rained most furiously. The cavalry returned during the night, and on the morning of the 21st we started back for New Berne, where we arrived about six o'clock, P. M. The roads, on our return, were as bad as they could be consistently, and we had considerable trouble in crossing corduroy roads, between the cross roads and Pollocksville. And when on a trot, at one time, the fore wheel of the gun carriage upon which Jerry McClair sat, broke through the corduroy; so suddenly was the carriage stopped, that Jerry was thrown from his seat sprawling into the ditch, and completely drenched with mud and water. Again, when near New Berne, in crossing a smooth level piece of ground, on a full trot, the wheel of the caisson, on which Wilbur M. Hoyt sat, struck a rut and threw him off in such a way, that his head lay so near the track that the hind wheel run over the cap he had on his head, and barely escaped the head, which must have been crushed had it been run over. When on our way to Trenton from Pollocksville, Major Frankle gave out strict orders against foraging or plundering, but Pierce Fitzpatrick, who was along, as an extra duty man, not knowing what his especial duty was, further than serving his country as he might perhaps have to do, and being provided with an extra horse, he conceived a plan whereby he might minister to the wants of the Twenty-fourthers, and make it pay also. He fancied that he was especially constituted for his plan, for he could keep one eye on the Provost Marshal and the other on chances to gobble. And that essential qualification, combined with business tact, he thought would carry him through safely, without any doubt whatever. He therefore provided himself with some paper and a pencil, and wrote receipts for money which he might pay or not, for geese, turkeys, or chickens, &c, and our receipt would cover any or all he would get, for had it been written in Greek it was equally as intelligible, but Pierce said it was as good as the natives could generally write, and if he should get caught, the receipt, as he interpreted it, would be good and satisfactory to the "Dutch Provost."

Accordingly, he sailed off on his steed, to the head of the column, and when the first chance offered he started for a large plantation house, and there gobbled, after some trouble, two geese, which he started for the company with. But just as he came out of the lane, who should he meet but the Provost, who accused him of plundering, and ordered him to put the geese into an ambulance, and fall in with the guard, under arrest, at the rear of the column. Pierce protested and produced his receipt, but the Provost avowed he was not going to be humbugged by him, and placed him, accordingly, under arrest; where he remained until the following day, when, at the instance of Captain Ransom, he was released.

When we arrived at Young's Cross Roads, being short of rations, which was reported at head-quarters, the Provost sent Pierce the geese, which had in the meantime spoiled. And there it was that Pierce's righteous indignation was fully aroused, and he d—n'd the Dutch Provost. Here, at Young's Cross Roads, we were not in camp more than ten minutes before we had two hogs killed, and well nigh dressed; but not wanting but one we gave the other to some infantry men, belonging to the Forty-third regiment; while taking it to their camp, they were arrested and placed in irons. At the same time nothing was done with our boys. We here reported being out of rations for our horses, and were accordingly granted permission to pass the picket guard, and when outside, we found an old bachelor who had a smoke house full of hams, and we accordingly filled our sacks with hams instead of corn, and brought them in, and had several in camp; when in passing the officer of the guard, coming in, one of the bags untied and let out the would-be corn (hams), and thus revealed the case, and put a stop to foraging in a hurry. We lived well on that march, and enjoyed it very much.

About January 26th, 1863, General Foster left New

Berne with a good share of the troops under his command, to co-operate with a fleet which sailed for Charleston, S. C., from Beaufort, N. C. As the Twenty-fourth Battery did not accompany this expedition, an account of it would not be in place in this book. It is enough for us to say, that on account of a misunderstanding between Generals Foster and Hunter, the land forces accomplished nothing at Port Royal, their place of destination. We believe this was termed Admiral Dupont's expedition. Gen. Foster returned with a portion of his troops, about the middle of February. During his absence all the camps had been incited to rivalry in beauty of appearance. Streets were laid out in the camps, and rows of evergreens were planted along the walks. Arbors were constructed in front of the tents, overspreading brick walks and oyster shell door-steps. The grounds were policed twice a week. Everything had an appearance of neatness, beauty and health.

New uniforms, shining brass, white gloves, blacked boots and salutes to every officer you met—were the order of the day.

On the 26th of February, a grand review of all the troops in the department was made by General Foster. In this display of tinsel, music, array of men, and the usually imposing sight produced by a large number of soldiers in line and column, this review was the feature of our ornamental service in New Berne.

On the morning of the 5th of March, two brigades of infantry, two or three sections of batteries and a half a dozen companies of cavalry were ordered out upon the Trent road. The first section of the Twenty-fourth Battery was sent to Newport Barracks by railroad. The following morning the section, together with the Fifty-first Massachusetts Infantry regiment and two companies of cavalry, started for Cedar Point, a landing directly opposite Swansboro, on the White Oak river.

Arriving at Cedar Point, we held that place until signals were given from the other side of the river, by the large force which had marched directly from New Berne, that Swansboro had been entered and only two rebel soldiers could be found.

The object of the expedition was supposed to have been to capture a couple of companies of rebel infantry and some cavalry, which was reported to have been lurking about in that region. A total of ten or twelve were captured. The rest had fled. The Battery returned on Tuesday.

On March 14th, Company A, of the Third N. Y. Cavalry and two companies of the Twenty-fifth Mass., who were occupying a picket post six miles from the city, on the Trent road, were attacked by quite a large force of rebels.

One of the cavalry boys was killed. They were obliged to retreat from their camp and take a position a few miles this side, at the Jackson house. Cavalry reinforcements were immediately sent to them, and infantry and artillery followed. The morning following the camp of the Ninety-second N. Y. Regiment, on the other side of the Neuse river were attacked by a brigade of rebel infantry and sixteen pieces of artillery. The Ninety-second stood it bravely amid a perfect shower of grape and shell. The gun boats came to their immediate assistance. The rebels then commenced shelling our gun boats, some coming within a few feet of the Dudley Buck, and some striking in the camp of the Twenty-fourth Mass. Our Battery was ordered down to the river shore, we were there in twenty minutes after receiving the order. The distance across the river being two miles and a half, we could not reach them with shell and were obliged to use solid shot. The Twenty-third arrived about half an hour afterwards. General Pettigrew, commanding the rebels, sent into the garrison of the Ninety-second three times for them to surrender, and after sending his compliments, and refusing to do so twice, the third time Colonel Anderson told him to go to h——l. Under the heavy fire of all our artillery he soon retreated. Our loss was three wounded. The rebel loss, six killed, twenty-two wounded, and twenty-five horses killed.

CHAPTER IX.

1863.

PLYMOUTH.

Soon after the feint, as described in the last chapter, was made upon New Berne, advices were received from Washington, N. C., and Plymouth, that an advance was being made upon those towns. A section of the Twenty-fourth New York Battery was immediately sent to Plymouth, and by the 1st of April the whole Battery had been shipped to the same place. To the members of the Battery this proved to be our destination for a much longer time than was thought of at the time of the removal. It was a small garrison, well defended, and so situated that, as afterwards was demonstrated, a small force could hold five times its strength and numbers at bay for a long time.

The houses in the town had been mostly deserted by their original inhabitants, and the larger ones were taken possession of by the troops for quarters.

It was a small village, situated on Roanoke River, and probably numbered in its palmy days fifteen hundred to two thousand inhabitants; streets regular and shade trees in abundance. Prior to our reaching this place, many of the buildings, during an attack upon it, had been burned to the ground. Aside from the desolated appearance that these ruins gave, the town was pleasant and pretty. The river furnished a variety of fish in great abundance, and the countrymen were allowed to bring in meats, eggs, green corn, poultry, honey, &c., in their respective seasons, while the negroes, who were settled in the inferior huts about the suburbs, were employed as cooks; so that, all in all, we were about as comfortable as soldiers could legitimately be permitted to be. The only objection to remaining in the place was a prevalence of miasmatic and intermittent fevers.

The siege of Washington, D. C., was of short duration —the rebels soon retreating; they were pursued a short distance towards Kinston, but no general battle took place.

Upon evidence of a permanent stay being given to the officers of the Battery, the members were busily employed at building stables, converting houses into quarters, and fitting up grounds for our park.

The guns were parked, and for a time we served as cavalry on short scouts into the surrounding country, searching for spies, guerrillas, traders and forage.

These expeditions were exciting and enjoyable, conducive to good health and the developing of muscle; they were, as a rule, successful, we seldom failing to bring in either prisoners, contrabands or contraband goods.

Being on an outpost, the commanding officer was thus enabled to keep himself pretty well informed of any movements of the enemy.

The garrison life at Plymouth was, as a whole, a pleasant experience. Our quarters were comfortable and all our corporeal wants well cared for. The occasional scout into the country furnished excitement, topics for conversation, and contraband goods from the deserted houses of rebels, which last added much to the comfort and adornment of our tents and barracks. Our drill no more than afforded proper healthful exercise. Our guard and camp duty was only sufficient to keep us in proper discipline. We were allowed the limits of the town, and had we been settled in the village of Perry, as strangers, we could not have been made more happy or comfortable. We have a letter, dated September 10, 1863, and, as it describes about the ordinary routine of duty at that garrison, we quote from it:

> The languor and idleness introduced into the human organization by the present state of the weather is indescribable and almost unendurable. With the tent canvas thrown open at both the front and rear of the tent, we seek to "raise a breeze" and dispel the intensely hot atmosphere that pervades even the shaded places. We almost wish that it were within the rules of propriety and the United States regulations to appear in Georgia cavalry uniform, viz., a palm-leaf hat and a pair of spurs.
> The uncomfortable sensation of being too warm is not the worst of it. It creates right down laziness, a disposition to seek a position "far niente," and to wear the time listlessly away. But I must write home. I have no subject to write upon, no tale to tell. We are doing nothing but grooming horses, polishing brass, oiling revolvers and scouring sabres. Afterwards, at our leisure, as the sun goes down and the atmosphere becomes cooler, we circulate through the town, mounted.
> In the evening we enjoy story telling, and have our musical sociables. We have a melodeon, guitar, violins and a flute—quite an orchestra, isn't it?
> Let me describe to you my day's labor, and I know you will laugh with me.
> C. went with his detachment on a scouting expedition up the river, on the steamer *Rucker*. As they started early, I rose at 5 A. M., and

prepared breakfast. Our culinary department is quite extensive; and as I have often informed you, so I repeat, I shall come home an accomplished cook. After they were off, I went out to the stable, and, in stable frock, flourished brush and comb with the dexterity of an accomplished groom. This season of the year causes grooming to be a most laborious task, I assure you, and the perspiration rolled off my face pretty freely. Then I took my saddle and bridle down to the artificer's to be repaired, but Rawson was sick, and therefore I was obliged to repair it myself. I know I sewed from me, and made big holes with the awl, and the job was rather bungling, but it was strong and answered the purpose. From there I came to my tent, put on the potatoes, drew my ration of steak and prepared it for dinner. Though alone, my appetite was strong enough to make me enjoy it greatly. I then cleared the table, washed the dishes, swept and dusted; took my new jacket, and with scissors, thread and needle, remodeled it to quite a genteel fit. This occupied so much of the afternoon, that I only had time to take a short galop out into the country and pick a few luscious persimmons. As I returned, the bugle was blowing the water call, and since then I have been quite busy, taking care of the horses, eating supper and preparing the night's wood. Our table furniture, by the way, is quite grand. We have china plates of different sizes, white-handled knives and forks, cut-glass tumblers, etcetera, *ad infinitum*.

Crooker, you know, has just returned from his home furlough. On the day he returned we asked him down to dine with us. We happened to have chicken pie that day, with our usual vegetables, bread, &c., all of which he appeared to highly relish and appreciate. As he finished, he pushed back his mustache with the napkin and said, "Well, boys, I do wish your anxious mothers might look in upon us and this dinner. The best answer I could make to all their inquiries about comfort and plenty would be to point at this and say, 'Look there!'"

Thus you see it is in a soldier's life—the brightest of bright sides one day, the darkest of dark sides on the next—extreme inactivity or extreme hard labor—luxury or hardships. We know not what the hour may bring forth. Rejoicing in comfort, we may be suddenly called out for a march, a scout or an attack.

We have given the whole of this letter, since we believe that all our comrades will recognize it as a correct description of the life at Plymouth, with the exception of perhaps a weekly call to go on a scout or a reconnoitre. Among the most prominent of these was the skirmish at Williamston. The following description of the march to and skirmish at Williamston, N. C., is given in a private letter, dated August 6th, 1863:

Sunday, July 26th, two sections of our Battery were ordered to report on the Washington road, at 11 o'clock. We did so and found the Eighty-fifth N. Y. and the One Hundred and First and One Hundred and Third Pa. Regiments already there. The Battery took the centre and in a short time the line was formed, and we started. The day was pleasant, the roads good, horses impatient and the boys fresh and jolly. We marched very leisurely along, taking the Washington road, as far as Nichols Mills. Here we turned off and crossed Ward's Creek. It was a difficult and dangerous place to cross with artillery, but all other bridges had been destroyed. We succeeded in crossing without any serious damage. This work was the hardest of the day. We advanced as far as Janesville (fourteen miles) and went into camp about 7 o'clock. During the day the sun was extremely hot. Several of the infantry were sun-struck, and a host of them were obliged to fall in the rear, so overpowerd by the heat, that it was impossible for them to keep up. Captain Cady told us to report in the lightest possible marching order. We followed his instructions a little too closely. We had no blankets, no lunch, nothing but a saddle for a pillow and an overcoat for a bed-blanket. C—— had purchased a chicken on the road; that roasted on a stick, together with a couple of hard crackers, and a cup of coffee made us a very good supper; we were very tired, so we soon made a bed under a tree and sound, dead sleep quickly took us away from " marching " realities. At five the following morning we rose, groomed our horses, and then went down to the river and had a fine and refreshing bath ; soon after the march was resumed. On account of the destruction of a bridge, we were obliged

to leave the main road, and travel some seven miles out of the way.
This was a swampy, muddy track, and with the hot sun pouring down.
This route made men and horses fret, sweat and tire. Finally, we
again struck the main road, and as we did so, we discovered some
rebel cavalry, who immediately "skedaddled." White flags were
hung out over every gate as we passed along. After marching
about ten miles, report came that we had reached the rebel picket
posts. Soon skirmish firing began, and was briskly carried on.
Captain Cady and Lieut. Dolbeer had gone ahead of the main forces
to reconnoitre, and some stray rebel bullets gave Dolbeer a pretty
close call. Immediately, however, the rebels retreated and crossed
still another stream. It began to rain, in fact to pour copiously. The
first two guns, Clark's and Crooker's (Williams commanding Crooker's
on account of his absence) and the Eighty-fifth N. Y. were sent forward. They went on a quarter of a mile, and coming into battery in
a corn field, commenced to shell. It rained so hard that it was impossible to "see" anything, but they "calculated." The rebels replied with muskets, sharply. The boys held up a little and the infantry
exchanged volleys. It sounded savagely although no one was hurt.
Just then one of the Colonel's aids came up and told our section to
come forward and take the left with the One Hundred and Third Pa.
As the officers were all in front, Merrill was obliged to take command.
We went with a rush. The boys were greatly elated ; we were soon
at the place appointed for us. The other boys had discovered a stone
mill with some rebels in it, and put three shot through it. Musketry
was sharp and plenty of it. The cavalry had made a charge and two
of them were wounded. The bridge over the stream had been torn
up ; the rain was pouring ; the stream swelling. The infantry had
managed to fire the saw mill which the rebels were using as barracks.
Just at this juncture the Colonel commanding (Col. Leymen,103d Pa.)
concluded to retire. How disappointed every body was. There is
nothing that causes a soldier to be more dejected or weary than to be
obliged to turn back after making an attack. The excitement was over
and the reaction came on in loud grumbles at the officer commanding.
We had felt confident of a victory, and were a good deal chagrined at
the idea of a retreat. We were obliged, however, to obey orders, and
at 8 o'clock, having had little or no dinner and no supper, we marched
back to our camp of the night previous, which we reached about

12 o'clock. During the march the rain poured constantly down, soaking every thing through, and at night we lay down under our paulins in the corner of fences, and on waking in the morning from sound slumber found ourselves lying in puddles of water. Resuming our retreat, after we had partaken of a little coffee, we began to feel the effects of a hot sun. The humid atmosphere seemed a cloud of hot steam, suffocating to each person as they breathed it. In the afternoon it again began to rain. A thunder storm, clouds low, and filled with electricity which seemed to follow our guns and musketry, enveloped us, and the terrific flashes of lightning and the deafening roar of the thunder, put to shame our artillery fires and reports.

Our march was through woods. The lightning seemed to play among the trees, now and then selecting some splendid pine and cleaving it to the roots, causing a clap and crash of thundering noise that made the very earth tremble. The rain was piercing; overcoats nor rubber coats were of any avail, the boys gave up any defence from the rain, and finally jumping from their seats marched through the stream that filled the road until we reached Plymouth. To some of the boys this was first experience, and to them it was pretty trying.

June 13th, Captain Lee, by special order, No. 168, was honorably discharged on surgeon's certificate.

E. H. Wardwell, while on a furlough home, had received a commission to fill the second lieutenancy which had been vacant for some time.

Soon after Captain Lee's return to the north, first lieutenant Cady received a commission as captain, second lieutenant F. S. Hastings was promoted to first lieutenant, and C. H. Dolbeer received a second lieutenant's commission. By order of Captain Cady, a reorganization of the Battery was made and the following is a roll of the names with the respective positions of the members of the Battery :

TWENTY-FOURTH NEW YORK BATTERY.

FIRST SECTION.
L. A. CADY, Captain.
George S. Hastings, First Lieutenant.

FIRST DETACHMENT.
W. W. Crooker, Sergeant.　　B. F. Corbin, Gunner.
Samuel Stoddard, Caisson Corporal.

T. Rich,	C. R. Griffith,
M. Crosby,	G. W. Kellogg.
P. Marrin,	A. Piper,
Charles W. Fitch,	P. J. Stafford,
H. Chadbourne,	Geo. Duryea,
A. Lee,	H. S. Whitney.

SECOND DETACHMENT.
C. A. Clark, Sergeant.　　Samuel Nichols, Gunner.
E. H. Hunter, Caisson Corporal.

W. Blood,	O. S. McCrary,
C. Wetmore,	T. McGuire,
C. T. Phelan,	G. Barker.
G. W. Keeney,	J. Filbin,
E. Richards, ♦	J. Russell,
W. Carnahan,	J. T. Ferrin,
E. Welch,	James Cowen.

SECOND SECTION.
C. H. Dolbeer, Second Lieutenant.

THIRD DETACHMENT.
R. C. Ainsworth, Sergeant.　　L. Newcomb, Gunner.
F. M. Alburty, Caisson Corporal.

W. Ainsworth,	J. E. Galusha,
M. C. Smith,	J. Flynn,
W. P. Nichols,	A. McDonald.
G. Miller,	J. G. Miner,
J. Sunfield,	H. Bartlett,
T. Fitzgerald,	J. H. Weller,

G. A. Holman.

FOURTH DETACHMENT.
J. W. Merrill, Sergeant.　　G. G. Wright, Gunner.
E. T. M. Hurlburt, Caisson Corporal.

L. H. Shank,	E. Eastwood,
W. F. Hosford,	R. J. Newton,
A. Lent,	G. Crounce,

B. F. Bachelder.

A. L. Culver,	P. Fitzpatrick,
C. A. Marean,	Le G. D. Rood,
J. Baker,	O. G. Parmlee,
C. H. Homan,	A. W. Comstock.

THIRD SECTION.
Frederick E. Hastings, First Lieutenant.

FIFTH DETACHMENT.

O. Williams, Sergeant.   A. T. Ferguson, Gunner.
G. B. Johnson, Caisson Corporal.

A. Griffith,   W. M. Hoyt,
P. Shirley,   J. Button,
L. H. Lapham,   H. V. Clnte.
F. Leonard,   Z. Allen,
G. W. Stevens,   M. R. Mosier,
Sylvanus King,   J. Bartley,
                 W. Gould.

SIXTH DETACHMENT.

William S. Camp, Sergeant.   Geo. Birdsall, Gunner.
H. Tilton, Caisson Corporal.

J. Woolsey,   H. Loomis,
G. S. Atwood,   P. Pratt,
                 J. W. Perkins.
J. McCrink,   C. Hathaway,
R. H. Barnes,   S. Root,
J. A. Brooks,   W. Armstrong,
                 E. Wood.

Edward H. Wardwell, Second Lieutenant, Chief of Caissons.
H. P. Lloyd, Orderly Sergeant.
H. C. Martin, Quartermaster Sergeant.

A. Richards,   }
P. D. Rawson,   } Artificers.
P. Calteaux,   }

H. C. Burd,   } Buglers.
W. A. Whitney,   }

W. Alburty, Guidon.
J. Chapman, Wagoner.

B. H. Hollister,   } Cooks.
J. Calkins,   }

Not long after the reorganization, and while we were busily engaged in making improvements in the appearance and the comfort of the camp, preparing our stables for the winter, &c., we were surprised by the arrival of General Butler—then commanding the department to which we belonged—on an inspection tour. We find a description in a private letter, from which we extract the following:

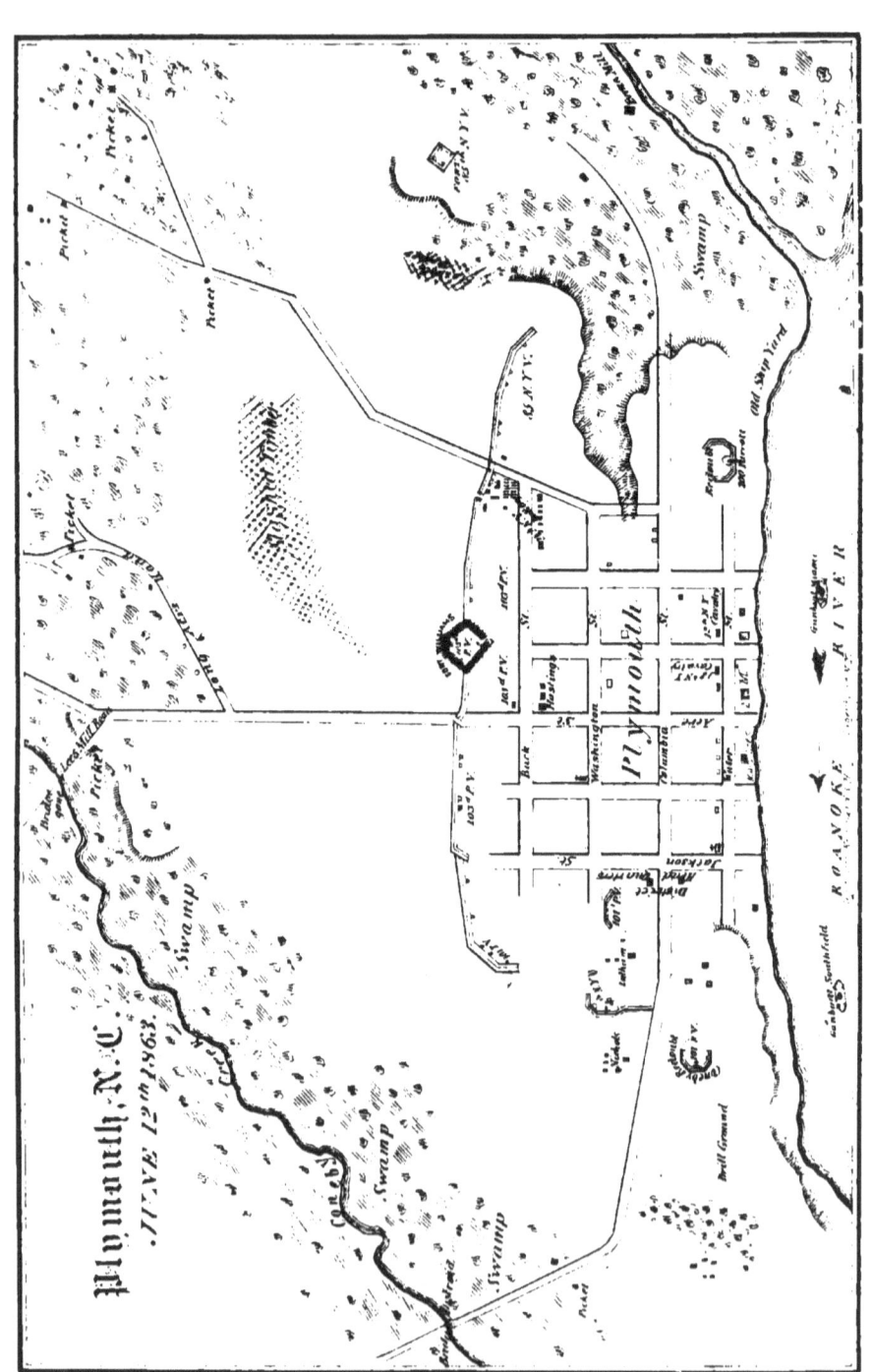

Yesterday Plymouth was alive, noisy, gay. In the early morning the steamer "Colyer" sailed up the Roanoke, with the colors of the General commanding. Immediately orders were issued—"General Butler has arrived, prepare for inspection"—"the Battery to fire a salute." The barn was not completed; nearly everything was out of order. There was much work done in short time. Blanks were made, guns scoured, grounds policed, harness cleaned, arms and equipments brushed and polished, boots blacked, clothes cleaned, and everything placed in "inspection" order. At eight o'clock we were hitched up and marched down to the parade ground in front of the Generals, and then we fired a salute of thirteen guns. We then returned to park. In about an hour General Butler arrived at our camp. He rode in a buggy with General Peck, driving his own horse. Reining up in front of the Battery, he alighted, and sportively assuming the part of coachman to General Peck, with an extremely low bow, assisted him in descending. Then approaching the Captain, with hat uplifted in acknowledgment of the "present" of the company, he shook hands with him, passed the compliments of the day, and requested him to dismount and walk through the Battery with him. The "inspection" was close; they observing all the minor as well as the more important things. Some of the boys in their hurry had forgotten to black their boots; *that* was noticed. Another man, whose pants were too long, had turned them up a little; the General allowed "he had got into the wrong man's pants." "The guns were in good order," he remarked once or twice. Both Generals said, "The men are fine looking, and their clothes in excellent condition." "Yes," says Butler, "well shod, well shod, too." General Wessells followed him, he in a buggy also, with Admiral Lee. They alighted, and came and examined our guns. Lee said "they were the best kind of field guns in use." They stayed with us from twenty minutes to half an hour; it was the most thorough and rigid inspection we have had.

Colder days began to come upon us, prophesying an approaching winter. We prepared for it, and were perfectly quiet during this season. One or two of the churches were put in order, and the chaplains of the different regiments alternately held service in them. There was quite an interest manifested among the soldiers on the subject of religion, and there was usually a large attendance at the services. It was somewhat singular that the preliminary attack on Plymouth was made while our men were returning from the church to their camps. A large "contraband" school was instituted, and held its sessions in one of the churches. As many as six hundred colored people, young and old, took advantage of this opportunity to study and learn. The school was superintended by Mrs. Freeman—a woman eminently fitted for the position—assisted by her daughter, and Mrs. Coombs, from Ohio. These ladies, when the approach of the enemy was known, were sent by steamer to Roanoke Island, where they finally established a very large and very successful colored school under the auspices of the "Christian Commission."

The sufferings which these ladies lessened among both the blacks and the whites, and the good they did, make for them a name that shall live for ever in the hearts of the poor creatures to whom they gave knowledge, comfort and consolation. In addition to our churches and schools, the gayer portion of the garrison interested themselves in concerts, balls and parties. "Ferguson's Band" was in as great demand as it is in the present winter seasons.

The few whites who were left, and able to entertain, did so. The officers of the different departments entertained and the soldiers did the same. Christmas was kept as a holiday, and on New Years, the day was made jolly by a show of climbing greased poles for a purse, running sack races, chasing a greased pig, running races with wheelbarrows while blindfolded; the whole concluding with a grand scrub race of all sorts and sizes of horses.

It was at this time that the older portion of the Battery, who had served their full term of service of three years, were induced to re-enlist.

A short time after their re-enlistment, they received veteran furloughs, and in a body left Plymouth for home.

On their return, Captain Cady again made a change in the roster of the Battery; many vacancies having occurred by resignations, promotions, deaths and sickness. We have no copy of this last roster. We remember, however, that Lieutenants Fred. E. Hastings and Dolbeer left the Battery. (We were told, at the time, that the reason of their departure was, that the number of the men in the Battery was too small to entitle it to so many commissioned officers.) William S. Camp was appointed quartermaster sergeant, and W. P. Crooker was appointed orderly sergeant. The duty sergeants were reduced in number, as were all the non-commissioned officers.

CHAPTER IX.

THE BATTLE OF PLYMOUTH.

Lieutenant George S. Hastings has kindly written for us the following sketch of the battle of Plymouth:

On Sunday, the 17th day of April, 1864, at the hour of dress parade, the pickets of the Plymouth garrison were driven in by the rebel cavalry forming the advance of the division which was rapidly advancing upon the post. The long roll was hastily sounded, and our troops hurriedly prepared for the attack. The garrison then consisted of the Eighty-fifth Regiment New York Infantry, the One Hundred and First and One Hundred and Third Regiments of Pennsylvania Volunteers, the Sixteenth Regiment of Connecticut troops, one company of the Twelfth New York Cavalry, two companies of the Second Massachusetts Heavy Artillery, and the Twenty-fourth New York Battery, numbering in all not more than 1,900 effective men. The rebel cavalry dismounted and deployed under cover of the woods, 1,200 yards from the outer face of our works. Our cavalry was quickly sent out to draw the enemy's fire and to discover their strength, and, when within range of the woods, received a sharp volley from the concealed rebels, which killed one man and severely wounded Lieutenant

Russell of the advance guard. It became evident, from the nature of the attack, that it was not simply a dashing raid, and our troops were accordingly prepared for the hard fighting soon to follow. Shortly after the demonstration upon our front, the shells of a rebel battery began to fall within our works. These first came from guns opening upon Fort Gray, a small but strong earthwork on the Roanoke River, about a mile from Plymouth, commanding the water approaches above us. This desultory fire, while doing little or no damage, was accepted by us as additional evidence of the seriousness of the attack. All night long the heavy music of artillery and the bustle of hostile preparation continued. About midnight the steamboat "Massasoit" left us, carrying to a safer point the "impedimenta" of the garrison, consisting of women, children and the disabled. The writer still retains in vivid remembrance the hasty farewells then and there spoken (some of which were final), the pale faces of affrighted women and children, the groans of the sick and wounded, and the bustle and confusion which, if reproduced, would form so striking and touching a picture of war. He well recollects how proudly the gallant Flusser (the lieutenant commander of the little fleet of gunboats guarding the waters of the Roanoke) paced the decks of the "Massasoit," with brave words like these, "Ladies, I have waited two long years for the rebel ram. The navy will do its duty. We shall sink, destroy or capture it, or find our graves in the Roanoke."

On the following day the enemy maintained a steady though ineffective fire upon our redoubts until evening, when they assumed a vigorous offensive. During the afternoon our skirmish lines had been busily employed in the dangerous exercise of giving and taking powder and ball. About five o'clock, however, the enemy advanced in force along our entire front. Our men fell back in excellent order, keeping up a sharp fire against the solid line of rebels. Behind this dense curtain of infantry, their artillery, consisting of about forty pieces, was advanced to a line about 800 yards from our outer works. Then rapidly taking position, their guns were served with terrific rapidity and precision. Our artillery responded deliberately and with fearful certainty, silencing the guns of the enemy and sending destruction into their ranks. So exact was our range, that in some instances a single shot disabled the rebel piece which had invited the salute. For nearly an hour this duel of artillery continued, the heavy ordnance of the gunboats adding their thunder tones to the chorus of death. As the fierceness of the attack subsided, the shrieks of the wounded and dying could distinctly be heard above the din of battle. The rebel infantry, which had been lying down during the heavy fire, must have suffered severely from our shells, and we believed that half at least of their artillerymen were put out of the fight. A rebel officer was heard to exclaim, " It is of no use ; we cannot endure this fire ;" and so their troops were withdrawn. Had the original design of carrying our lines been further pursued, we were confident that canister and the bayonet in closer quarters would have proved too much for the mettle of the Southern veterans. Their attack was well planned, and would doubtless have succeeded, but for the strength of our earthworks, which protected us from a fire that would otherwise have been most damaging. As it was, our casualties were comparatively light, though the air was full of the missiles of death. During this formidable demonstration against our lines, a strong storming party, under the command of Colonel Mercer, of Virginia, attempted to capture a small redoubt, which by some freak of engineering had been located nearly half a mile from the main defences. This redoubt was occupied by a single company of the Eighty-fifth New York Volunteers, commanded by Captain Chapin. Again and again the rebels charged upon this little garrison, coming within range of their hand grenades. Their reception was so warm, that they too were compelled to retire, leaving some thirty or forty prisoners in the hands of the brave defenders of the little fort. The brigade commander, Colonel Mercer, was killed in the assault.

Thus far, the Plymouth troops were confident of repelling the enemy. Later in the night, the rebel ram "Albemarle", succeeded in passing our batteries and sinking two of the gunboats, inflicting a very serious loss upon our naval forces. The gallant Flusser, while holding the lanyard of one of his guns, was struck by a piece of a shell and instantly killed. This reverse, and the consequent withdrawal of our naval supports, and the undisturbed occupancy of the river by the rebels, gave a serious phase to the siege, and our capture then seemed to be a question of time and endurance only. Our left was now no longer protected by the powerful batteries of our gunboats, but exposed to a galling fire from the "Albemarle" and her wooden convoy. Our troops then commenced throwing up bomb proofs, as a protection from the fire in the rear. The rebels moved their artillery and infantry to our left, which was plainly our weakest point. Continuing their fire upon our front and right, the bulk of their forces was quietly moved into the fresh position. About midnight of Tuesday, April 19th, in the teeth of a sharp and destructive fire, they laid their pontoons across a creek intersecting the open ground lying just east of our left line. Crossing with two brigades of infantry and several pieces of artillery, they formed a new and strong line of battle, the right of which rested upon the Roanoke and the left swerving around to our front. At the same time, another force advanced against our right line. About three o'clock, on the morning of April 20th, the entire rebel force charged our extended and feeble lines, moving forward with loud and defiant yells. Largely outnumbering our exhausted garrison, they were able to make a vigorous onset upon every portion of the defences, and at the same time to send an independent column along the banks of the river into the heart of the town. This final success was achieved with great losses upon both sides. The pieces of the Twenty-fourth New York Battery were served double shotted with canister, hurling disorder and death into the ranks of the enemy; and not until the rebels seized the muzzles of the guns, did the cannoneers fail in their work. For nearly two hours did the fight go on in the streets of Plymouth, our forces surrendering only under stern military necessity and in small detachments. Fort Williams turned its guns upon the rebels, and did murderous execution for three or four hours.

Finally, when every portion of that strong earthwork was covered by rebel sharpshooters, and the rebel artillery had been so disposed as to send a concentric shower of shell within its parapets, General Wessells accepted the situation, and saved the garrison from certain sacrifice by a reluctant surrender. The rebels raised the black flag against the few negroes found in uniform, and mercilessly shot them down. Their losses were never accurately known, but were stated in the Raleigh papers as exceeding 2,200 in killed, wounded and missing. Measured by results, the victory, so dearly won, was barren, as Plymouth was a point of little strategic value. The subsequent movements of the rebel forces showed the ulterior design of driving the Union troops from the State. This cherished plan would probably have succeeded, had not the movements of the army of the James caused the hasty recall of the division in North Carolina. Viewed in any light, the battle of Plymouth afforded a splendid illustration of the valor and sterling qualities of the American soldier. The rebels showed a stout tenacity of purpose and a courage worthy of a better cause. But in the 1,900 defenders of the post, they found men equally dutiful and brave. Our losses in killed and wounded were over 180—a fact sufficiently attesting the heroic conduct of our men, when it is recollected that, for the most part, they fought under cover of strong breastworks.

We give in addition to Lieutenant Hastings sketch, the following selections from the account written at the time by the correspondent of the New York Herald,

and General Peck's official report as found in the
"Rebellion Record:"

NEW BERNE, April 21, 1864.

General Wessells, commanding the town of Plymouth, and his whole command of upwards of two thousand, officers and men, surrendered yesterday at one o'clock P. M. The command consisted of the following:

Eighty-fifth New York Infantry Regiment.
One Hundred and First Pennsylvania Infantry Regiment.
One Hundred and Third Pennsylvania Infantry Regiment.
Sixteenth Connecticut Infantry Regiment.
Two companies of the Massachusetts Heavy Artillery.
Two companies of the Second North Carolina Volunteers.
Two companies of the Twelfth New York Cavalry.
Twenty-fourth New York Battery, 100 men, 6 guns.

The enemy gained likewise upwards of thirty pieces of artillery of all calibres, including one two hundred and one one hundred pounder Parrotts, about three hundred horses and a large amount of commissary stores.

FORTRESS MONROE, April 24, 1864.

PLYMOUTH SURRENDERED

is the startling and painful announcement I am compelled to make to you in my despatch to-day, which event took everybody by surprise, as it was thought that General Wessells could hold out for a few days at least, until reinforcements, which were already on the way, could reach him. But the rebel ram which had destroyed the Southfield kept our transports from ascending the Roanoke River, and consequently the beleaguered garrison at Plymouth was compelled to fight as long as human endurance could stand it, and either be annihilated or surrender at discretion. This news reached me this morning by the arrival of the steamer Currituck from Roanoke Island, and through a most reliable source.

THE FIRST ATTACK

was made on the fortifications of Plymouth on the 17th inst., but repulsed, as also another made on Fort Gray. The momentary repulse kept the enemy at bay, and lasted for nearly twenty-four hours.

On Tuesday morning the rebel ram made her appearance, to co-operate with the land forces, and succeeded not only in sinking the Southfield, but in killing Captain Flusser. They drove our naval vessels from the river, and consequently this support failed General Wessells in the most trying hour of his campaign.

THE SECOND ATTACK.

was made by the enemy on all of our forts surrounding the town almost simultaneously, and in every assault he was driven back with terrific slaughter. But the rebels seemed determined, and renewed attacks were made and checked, each one still attended with the most dreadful carnage. From this time until half-past ten o'clock on Wednesday morning the fighting was almost uninterrupted. On Tuesday orders were issued for

THE EVACUATION OF FORT WESSELLS,

better known as the Eighty-fifth Regiment redoubt, situated a short distance from Mill Creek. Captain Chapin, the commandant of the fort had been killed, and although the stock of ammunition on hand was growing "small by degrees and beautifully less," still the heroic little band kept up an incessant firing on the enemy, every shot telling with fearful effect. Upon the reception of the order the survivors of the garrison awaited the coming of night, and under its protecting shade silently withdrew to the town, having first, however, disabled the guns of the fort.

THE UNION FORCES CONCENTRATED,

on the evacuation of Fort Wessells, in Forts Williams, Comfort, and a redoubt facing the Roanoke River, below Mill Creek, and kept up the fight incessantly. The forces of General Wessells thus being massed and able to handle the artillery more effectively, kept up a regular, steady and galling fire on the rebels. The enemy used thirty-pounder Parrott guns and other artillery of about similar calibre.

FORT WILLIAMS STORMED.

At nine o'clock on the 20th inst. a most impetuous assault was made by the rebels on Fort Williams. Our brave boys nobly stood by their guns, and repulsed every attempt of the enemy to enter the fortification. In splendid order did the rebel column advance to the assault. General Wessells allowed him time to come within easy range of his guns, and then gave the order to fire. Every discharge mowed down the rebel troops by platoons. Still the gaps were instantly filled up and the attack renewed. In this manner the enemy received several severe shocks, and after a last and still more impetuous charge, which likewise resulted disastrously to him, he withdrew, evidently to repair damage and make ready for another attack.

GENERAL WESSELLS CAPITULATED

at eleven o'clock on the morning of the 20th inst. (Wednesday), an hour and a half after the repulse at Fort Williams. At the hour above mentioned the Union flag was hauled down on Forts Williams and Comfort, as well as on the Mill Creek redoubt. The rebels had been heavily reinforced during Tuesday night, and the overwhelming forces hurled against our weak and already shattered column was too much to endure, and being out of ammunition and cut off from a further supply, General Wessells could do no less than surrender, or have all his command annihilated. The garrison of Fort Gray no doubt fought nobly to the last ; but, being cut off from the main command, without hope of deliverance, had to succumb also. We have no advices from this post, but common sense teaches us that the fort could not hold out very long.

THE REBEL COLUMN

consisted of no less than five brigades of troops, each brigade numbering about three thousand men. These were under the chief command of Major General Hoke, assisted by Generals Ransom and Barton. The majority of these troops were from the far South, as the North Carolinians are not trusted very far while fighting on their own soil.

OUR LOSSES

are estimated at about one hundred in killed and wounded. Captain Chapin was killed at Fort Wessells, and Captain Horace J. Hodges, Depot Quartermaster, while in the act of communicating with the gunboat Miami, in a canoe which he carried in a wagon from Plymouth to Conesby Creek, in order to elude the rebel iron clad, was upset and drowned.

THE REBEL LOSSES

are, beyond the slightest doubt, immensely heavy, when it is considered that every fort around Plymouth was stormed from three to seven times, and each assault repulsed with great slaughter, besides pour.

ing broadside after broadside into the rebel ranks from the Miami and Southfield, the casualties among the rebel troops must have been enormous. A rebel surgeon was heard to say that "the damned Yankees had killed and wounded one-third of their whole force, and he hoped that no mercy would be shown the cursed Yankees." The gunboat Whitehead went on a reconnoissance on Wednesday, a short distance above Plymouth, and the officers and crew observed about three hundred rebel troops engaged in burying the dead. From a steeple on the town church, overlooking a large tract of land, it was found that the field of Asa Johnson (about sixty acres) was completely filled with dead and dying rebels. The entire rebel force could not have been short of from fifteen to twenty thousand men, of whom one-third are unfit for future service.

GENERAL PECK'S OFFICIAL ANNOUNCEMENT OF THE SURRENDER.

GENERAL ORDERS—NO. 66.

HEADQUARTERS OF THE ARMY AND
DISTRICT OF NORTH CAROLINA,
NEW BERNE, N. C., April 21, 1864.

With feelings of the deepest sorrow, the Commanding General announces the fall of Plymouth, N. C., and the capture of its gallant commander, Brigadier General H. W. Wessells, and his command. This result, however, was not obtained until after the most gallant and determined resistance had been made. Five times the enemy stormed the lines of the General, and as many times were they handsomely repulsed with great slaughter, and but for the powerful assistance of the rebel iron clad ram, and the floating sharpshooter battery, the Cotton Plant, Plymouth would still have been in our hands.

For their noble defence the gallant General Wessells and his brave band have and deserve the warmest thanks of the whole country, while all will sympathize with them in their misfortune.

To the officers and men of the navy, the Commanding General tenders his thanks for their hearty co-operation with the army, and the bravery, determination and courage that marked their part of the unequal contest. With sorrow he records the death of the noble sailor and gallant patriot, Lieutenant Commander C. W. Flusser,

United States Navy, who, in the heat of battle, fell dead on the deck of his ship, with the lanyard of his gun in his hand.

The Commanding General believes that these misfortunes will tend not to discourage, but to nerve the Army of North Carolina to equal deeds of bravery and gallantry hereafter.

Until further orders the headquarters of the sub-district of the Albemarle will be at Roanoke Island. The command devolves upon Colonel D. W. Wardrop, of the Ninety-ninth New York Infantry.

By command of   Major General JOHN G. PECK.

J. A. JUDSON, Assistant Adjutant General.

THE PRESS DESPATCHES.

NEW BERNE, N. C., April 22, 1864.

The battle, which had been going on night and day at Plymouth, from Sunday, the 17th, to the 20th inst., resulted in the capture of the city by the enemy on Wednesday noon, including General Wessells and his forces—one thousand five hundred men. The enemy obtained possession of the town at eight o'clock in the morning.

General Wessells and his troops retired into Fort Williams, and held out until noon, repulsing the enemy in seven desperate assaults. The enemy's loss is said to be one thousand seven hundred, while our loss was slight.

General Wessells, who gained such distinction in the seven day's fight before Richmond, has made in this siege a most heroic resistance with his little band of veterans. Several weeks since he called for five thousand men, stating in the most solemn manner that it would be impossible to hold the city with a less number. General Peck, who has given General Wessells all the assistance in his power, in the same solemn manner, time and again, called for reinforcements.

CAPTURE OF PLYMOUTH, N. C.

HEADQUARTERS ARMY AND DISTRICT OF NORTH CAROLINA, NEW BERNE, N. C., April 25, 1864.

General: I have the honor to submit the following report upon the loss of Plymouth, which is as full as it can be until General Wessells is able to make his reports, when I will make a supplementary one:

On the twentieth, at seven o'clock P. M., I received your communication of the seventeenth, in reply to the letter of General Wessells of the thirteenth, asking for reinforcements. As this letter must have reached your headquarters in the evening of the fourteenth, or early on the fifteenth, a reply could have reached me on the sixteenth, in time to have communicated with General Wessells during the evening or night of the seventeenth.

Unfortunately, the reply was not written until the seventeenth, and did not arrive on the twentieth until some hours after the fall of Plymouth.

You replied, viz.: " You will have to defend the district with your present force, and you will make such disposition of them as will, in your judgment, best subserve this end."

General Wessells sent his communication direct to your headquarters, to save time, expecting that any aid which might be sent would come from Virginia, and not North Carolina.

He sent a duplicate to me, with a letter expressing the above views, knowing the reduced state of the force at my disposal. He writes, viz.: " I have no idea of getting any troops, but have always been anxious to see more troops in North Carolina."

Notwithstanding this expression of his sentiments, I had a conference with General Palmer and Commander Davenport, United States Navy, and the heavy gunboat " Tacony," which is equal to two or three regiments, was immediately despatched to Plymouth.

On the eighteenth instant the " Tacony" arrived back from Plymouth, with despatches from General Wessells and Commander Flusser.

General Wessells wrote that he did not apprehend any attack, and did not think there was a large force in his front.

He expressed the opinion that there was doubt as to the " iron-clad" making its appearance, and believed that she was at Hamilton, undergoing repairs or modification.

He wrote on the sixteenth, viz.: " I have the honor to report that the gunboat ' Tacony' arrived here to-day, but, as her presence at this time does not seem to be necessary, I have so informed her commander, and he proposes to return to New Berne to-morrow. I cannot learn that there is any considerable force of the enemy on the river now, though such is the report from various sources. I very much doubt if there is any design of bringing the thing (iron-clad) down ;

still there may be, as they say, when the 'Neuse Raur' is ready. I am desirous of seeing more troops in this State," &c.

Commander Flusser also wrote to Commander Davenport, Senior Naval Officer, viz.: "I think General Peck misinterpreted General Wessells letter. We have had no scare here yet, and not even a small one for several days."

These able commanders had so much confidence in their ability to hold their positions against them, that they sent back the reinforcements sent them. This action placed me entirely at rest respecting affairs at Plymouth.

On Monday (eighteenth) afternoon, about 5:30 P. M., I received advices by deserters that General Corse was in front of the outposts at Bachelor's Creek with a large force of all arms, and that General Pickett would attack Little Washington on Tuesday. This information, taken in connection with that from General Wessells of the sixteenth instant, respecting the disappearance or diminution of the force in his front, led the authorities here to believe that Little Washington would be attacked immediately. Two steamers, loaded with troops, together with the gunboat "Tacony," were despatched to Little Washington. At an early hour on Tuesday morning, the nineteenth instant, despatches were received from General Wessells and Commander Flusser, announcing an attack by rebel land force on the afternoon of the seventeenth instant. This was the first information received from General Wessells subsequent to the sixteenth instant, when the "Tacony" was back, as above stated. The latest information received, through a contraband, the servant of Captain Stewart, A. A. General, General Wessells staff, is to the effect, that early on Tuesday morning the "iron clad" had complete control of the Roanoke River, and, in conjunction with the floating iron battery—the "Cotton Plant"—was attacking the town in rear, while the land forces were engaging our troops in front. From this statement it will be seen that the enemy had complete control of the Roanoke River within a very few hours of the time I received General Wessells despatch of Sunday night, the seventeenth instant. On the reception of these despatches, which were very favorable, steamers were despatched with such available infantry as General Palmer could spare,.together with supplies of ammunition for the army and navy at Plymouth. These steamers were detained in the Albemarle Sound by the gun-

ANDERSONVILLE HOSPITAL.

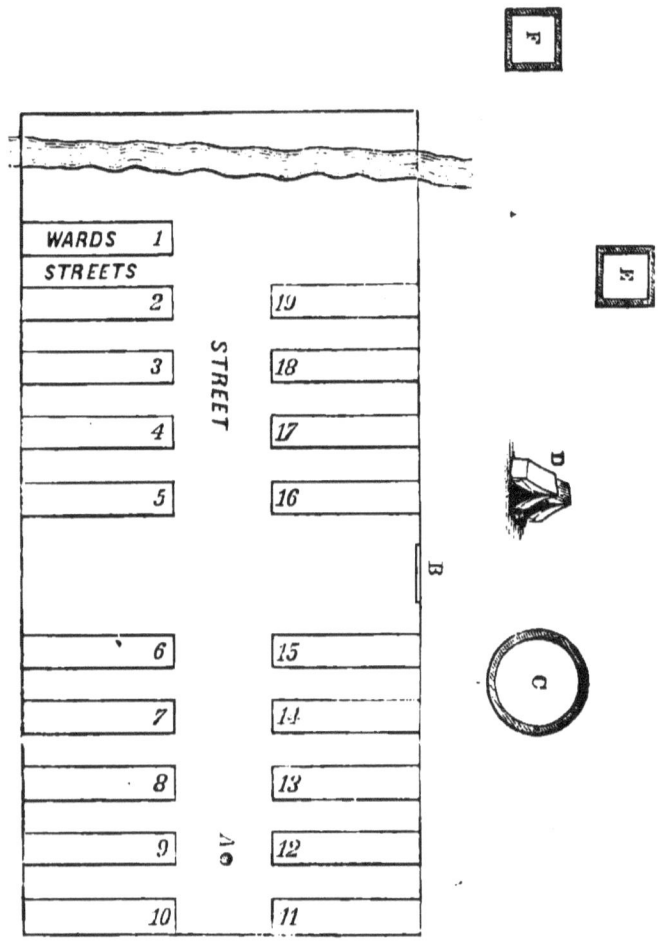

A WELL.
B GATE.*
C FORT.
D HEAD-QUARTERS TENTS.
E DISPENSATORY.
F DISSECTING HOUSE.

*The other sides of the Hospital border upon Swamps.

boats then lying in Edenton Bay, which had escaped from the "ironclad" at Plymouth. In my judgment, the non-arrival of the infantry at Plymouth is most fortunate, as they, together with the steamers, beyond doubt would have fallen into the hands of the enemy.

A steamer, with despatches, was promptly sent to General Harland, commanding at Little Washington, notifying him of the state of affairs at Plymouth. He was also requested to send down such surplus troops as he might have, to be used at such points as might seem necessary. I also sent despatches for the "Tacony" to proceed at once from Little Washington to Plymouth. Before these despatches arrived, Colonel Dutton, Chief of my Staff, had procured the sailing of the "Tacony" for Plymouth, going on board himself. Colonel Dutton also suggested to General Harland that he should send the steamer "Pilot Boy" with the Seventeenth Massachusetts Volunteers to Plymouth, but General Harland did not feel at liberty to do so, in view of his situation.

All the information received by both the Senior Naval Officer, Commander Davenport and myself was promptly sent to your headquarters by both lines of communication. General Wessells was supplied with provisions, forage, ammunition, tools and other requisites for a protracted siege. His command numbered some two thousand five hundred at Plymouth, and the casualties were very small, notwithstanding five assaults on Monday. His position was intact up to the appearance of the "iron-clads" and "Cotton Plant" at three A. M., the nineteenth; and but for the loss of the river, he could have held the land forces at bay for weeks.

General Wessells and his command, and the navy, under the late gallant Lieutenant Commander Flusser, made a heroic fight, worthy of our arms. They deserve well of the country, and history will record in glowing terms their honorable conduct.

JOHN G. PECK, Major-General.

To Major-Gen. B. F. BUTLER, Commanding.

CHAPTER X.

1864.

CAPTURE OF PLYMOUTH.

Up to this time the history of the Battery had been a pleasant one. We had had our petty quarrels and animosities. We had suffered from jealousies and disappointments. We had accused and been accused. But these things were trifles after all, and it needed but a day of genuine trouble in common to bring us all to a common sympathy, and an interchange of kindly words and kindly feelings. So far, I say, our army experience had been unusually free from hardship. In Newport Barracks we had comfortable log houses for our quarters, plenty of food, plenty of clothing, and only enough of drill for good gymnastic exercise. In New Berne we had been furnished new tents and new barracks, and there too our quarters, food and raiment were excellent and our duty comparatively light.

In Plymouth we used unoccupied houses for quarters, our scouting duty had been simply a pleasant excitement. The only affliction we had experienced was the monotony and the ennui arising from garrison duty. An inactive soldier's life is a lazy life at the best, and ignorant and unthoughtful of what the result might be, we welcomed the attack which had ended in bringing us all together as "prisoners of war."

There is no question that the defence of Plymouth by its garrison of 1,900 men, against a besieging force of 12,000 men, was one of the bravest and hardest fought battles of the war. The number of men on both sides were inferior to the vast armies contending in Virginia, but their determination and their courage could not be surpassed.

The defence and capture of Fort Chapin was a hand to hand fight of sixty men against two regiments. There was no surrender. The little band (two companies of the Eighty-fifth New York) held their own until their captain was disemboweled, their lieutenant wounded, and many of their comrades dead and wounded, and the few left could not prevent the rebels from clambering over the parapet and fairly crowding them out.

The writer was an eye witness of the combat between the rebel ram "Albemarle" and our gunboats, and believes that had the gunner in charge of the one hundred pound gun done his duty, by firing at her as she passed his fort, we never would have been captured. As she ran her prow into the gunboat Southfield, there were quick and loud commands on both vessels. The men of the ram were ready and using their small arms. The men of the Southfield had been taken by surprise, but were none the less active. I saw them trying to throw shells down the smoke pipe of the ram. They were also using hand torpedoes, but none had effect. The commander of our fleet was killed almost instantaneously, with the collision of the boats, and the captain of the accompanying gunboat, which had been chained to the Southfield, cast off, and steamed down the river. The success of the ram was the turning of the scale to the Confederates. With no fears from the raking range of the gunboats, they, on the morning of the 20th of April, at about three o'clock, made an assault with their full force, and with only our thin single rank at the parapet to oppose their overwhelming numbers, they drove our boys on the left through the streets of the town, killing and capturing them. Lieutenant Hastings was taken prisoner while galloping with one of his detachments to the assistance of the left flank, and found too late that they had been flanked, and that the troops approaching him were not the Union troops retreating, but the rebel troops advancing upon him.

On the right flank, while we were firing at the rebels in our front, we were surprised to see their gray coats at our right and rear. We double shotted with canister and fired, having the satisfaction of seeing many of them fall. A moment after we were surrounded and captured. We had, however, spiked our pieces and killed most of our horses.

On they went towards the center fort (Fort Williams), capturing on the way Crooker's section of our Battery, which had been doing good service at dismounting some rebel rifle pieces near Fort Williams, and scattering some rebel troops; and still further on, taking another section of our Battery under Captain Cady, which had been protecting the front of our fortifications. Finally Fort Williams was surrounded. We looked upon its defence as hopeless, but it was gratifying to our pride to see our commanding general (General Wessells) show so much pluck. He had fought his battle well. He had had no base to fall back upon, his disposition of his little handful of troops had been skillful, and yet it was hard for him to surrender.

All loyal citizens of the United States have a pride in our beautiful national banner, and ever is it a pleasure to their hearts to see it fluttering to the breeze. As children we learn to love it, honor it and cherish it.

Two epochs in my life have been strongly marked by the sight of this "emblem of the free." *First*—when it was slowly lowered from the color staff of Fort Williams at Plymouth, and the Confederate colors replaced it. *Second*—when for the first time in seven months I saw it waving from the masts of the vessels that had come to take us from our horrid prison pens.

In experiencing the first, it was a sad sight to see our pride, our boasted "Stars and Stripes" falling. We had fought for them, many of our comrades had died for them; but all was lost! Few of the many Union soldiers that stood around me had dry eyes as those colors fell. The future had no place in our thoughts, but the present made us vow that once again free from these cursed bonds, we would, stronger than ever, fight those men that dared pollute with their hands our flag.

Stripped of arms, mortified and sick at heart, we were penned by rebel guards, and allowed to take a night's rest on the green sward.

As the sun lowered we took a view of our once pleasant and happy camp; how desolate and dreary was it now. Proud in our own strength, we had been conquered.

How much of passion, hate and revenge rankled in the bosoms of even those who would be Christians. Our comrades killed, the battle lost to us, our friends at home frightened, anxious and full of sorrow ; our prospects for freedom from this degrading imprisonment far in the dim, dim future. Cruel taunts were thrown in our faces, cruel acts were committed on every side of us. We tried to brave it out, we tried to comfort ourselves with the knowledge that we had fought a good fight, we endeavored to believe that an immediate exchange of prisoners would take place, we consoled ourselves with the thought that none but cowards would taunt a fallen foe; yet heavy hearts and sad, sad minds dwelt with us all that long night.

The early morning found the rebels plundering and pillaging the town. Remarkable tastes were displayed by different men among the rebel soldiers, in selecting articles which they individually considered to be of the greatest value.

At ten o'clock on the day following the capture, we were ordered into line, and escorted by the oddest looking set of guards that a person could imagine. They were loaded down with dry goods, groceries, hardware, tinware, toys, clothing, bedding, woodenware, in fact you might say they had put the entire personal property of the village on their backs, and were marching off with it. We marched seventeen miles that day. If we may judge by the property strewn along the line of march, we think the rebels had the worst of that day's work.

The following day we marched ten miles. The next day we reached Hamilton.

On the 25th of April we reached Tarboro. At this place the officers who were prisoners were separated from the soldiers, and took the cars for Richmond.

The Union soldiers were divided into squads, and as fast as cars could be sent were shipped on platform cars for some Southern prison camp. The weather was extremely warm, and the only time any kind of comfort could be experienced was when the cars were in motion.

The first prominent place we reached was Wilmington. There was little sympathy tendered us there, since a squad of prisoners who had been through there just before us, had fired large quantities of cotton which was lying upon the wharves, and their fire department had been unable to control the flames. An immense sum of money must have been lost.

At this place there was a great deal of anxiety to purchase watches, jewelry or greenbacks for Confederate scrip. It seemed odd enough to be offered *one thousand* dollars for an ordinary silver watch, but at that time Confederate money was on a rapid decline.

The next day we reached Charleston. Here we received a great deal of kindness, and many tokens of sympathy. Water was given to us by women. Cigars, food, fruit and bouquets were handed to us by colored servants, with the compliments of their "massas" and "missus." Bouquets were thrown from windows to us. Words of encouragement and of condolence reached us in many ways.

From Charleston we were taken to Savannah, thence to Macon, and thence to Andersonville.

This then was our unknown destination.

It was quite dark before we were allowed to disembark from the cars. The stockade was about half a mile distant from the depot. We were told that before entering the prison we would be organized into detachments. We were marched to a level plot of ground, through which ran the stream which furnished the prison stockade with water; and after a guard had been stationed about us, we were permitted to furnish ourselves with water and appease our hunger with the bacon and hardtack that had been issued to us a couple of days before. That was the last of hard bread that I ever saw in the Confederacy. And here was my introduction to Captain Wirz. Camp fires had been started about the guard line; and suddenly, as if it had been the Devil himself, this fiend made his appearance through or near one of the fires. Short in stature, stooping figure, ill-shaped head, awkward limbs and movement, a deep-set, ugly eye, and a tongue reeking in profanity—such was Captain Wirz. A glance passed from comrade to comrade, telling better than the tongue of the fate we feared was in store for us.

After much swearing, and many threats to punish or kill, he succeeded in properly organizing us into detachments, and we were then informed that our barracks for the night would be the ground. Had we known then what was to be our future camping place, how quickly would our complaints have changed to words and thoughts of thanks—a practical example of the little we know in this world of the good or the bad that may be in store for us; while, in our ignorance, we are merry when we should be sad, and are full of complaints when we should be happy. Fatigue makes a soft, warm bed of the cold earth, and changes a stick of wood into a downy pillow. We slept soundly; and what a blessing, it would seem, it would have been had the great majority of our fellows never waked from that sleep. Still, Providence, wise and good, saw fit for them to wake, and to enter a trial of life that they had never anticipated. From observations in constant and intimate relations with many of them, I believe that long suffering and constant thought of the past and future did prepare them for a peaceful death and, I hope, for a blessed future.

On the following morning we were ordered into line, and marched into the prison stockade. It then contained about 10,000 prisoners, in an enclosure of five acres. As we moved through the gate, we were greeted upon every side by the inmates with salutations of sorrow and satire—eagerness for news—a great desire to "swap" corn cake for hard-tack—and a general disposition to make acquaintance with the new comers and their chattels.

The appearance of the place and its inmates was sickening, and our spirits drooped and hearts failed us, as our eyes wandered over the groups of ragged, swarthy, filthy, emaciated forms that grouped around us. I quote from the diary of R. Barnes, Sunday, May 1, 1864: "The prisoners look rough; I never see such a nasty place in all my life; we stay right out doors all night." The weather was exceedingly warm. We had no protection from the sun during the day, nor the dew during the night. The soil was sandy and full of fleas. The wood used about and in the stockade was mostly pitch pine, and the lampblack soot made by it settled upon the camp and the men, so that they resembled a delegation of unwashed charcoal men.

The stream of water was entirely inadequate for bathing purposes, and in a few days the brightest uniforms and the tidiest of our fellows began to bear near semblance to the oldest residents.

As this was our last camp, and proved to be to many of our dear comrades their last earthly abiding place, we think this a proper place to give a brief description of it.

CHAPTER XI.

ANDERSONVILLE.

After this, whenever any man who has lain a prisoner within the stockade of Andersonville, would tell you of his sufferings, how he fainted, scorched, drenched, hungered, sickened, was scoffed, scourged, hunted and persecuted, though the tale be long and twice told, as you would have your own wrongs appreciated, your own woes pitied, your own cries for mercy heard, I charge you listen and believe him. However definitely he may have spoken, know that he has not told you all. However strongly he may have outlined, or deeply he may have colored his picture, know that the reality calls for a better light, and a nearer view than your clouded, distant gaze will ever get. And your sympathies need not be confined to Andersonville, while similar horrors glared in the sunny light, and spotted the flower-girt garden fields of that whole desperate, misguided and bewildered people. Wherever stretched the form of a Union prisoner, there rose the signal for cruelty and the cry of agony, and there, day by day, grew the skeleton graves of the nameless dead.

But, braving and enduring all this, some thousands have returned to you. And you will bear with me, and these noble men will pardon me, while, in conclusion I speak one word of them.

The unparalleled severities of our four year's campaigns have told upon the constitutional strength even of the fortunate soldier, who alone marched to the music of the Union, and slept only beneath the folds of the flag for which he fought. But they whom fickle fortune left to crouch at the foot of the shadowless palmetto, and listen to the hissing of the serpent, drank still deeper of the unhealthful draught. These men bear with them the seeds of disease and death, sown in that fatal clime, and ripening for an early harvest. With occasional exceptions, they will prove to be short-lived and enfeebled men, and whether they ask it or not, will deserve at your hands no ordinary share of kindly consideration. The survivor of a rebel prison has endured and suffered what you never can, and what I pray God your children never may. With less of strength, and more of sad and bitter memories, he is with you now to earn the food so long denied him. If he ask "leave to toil," give it him before it is too late; if he need kindness and encouragement bestow them freely, while you may; if he seek charity at your hands, remember that "the poor you have always with you," but him you have not always, and withhold it not. If hereafter you find them making organized effort to provide for the widow and orphan of the Union prisoner, remember that it grows out of the heart sympathy which clusters around the memories of the comrades who perished at their side, and a well-grounded apprehension for the future of their own, and aid them.

CLARA BARTON.

Andersonville, Georgia, was a wood and water railroad station. It was located within nine miles of Americus, and for a time the prison encampment was designated as being located at Americus.

It was selected by the rebel authorities as a proper location for a military prison, since it was then nearly central as regarded the Confederate States, and their then probability of maintaining the ground held by them.

That portion, too, of the Confederacy was better able to furnish provisions and other supplies, having been quite remote from active scenes in the war.

The rebel camp of guard at Andersonville was called "Camp Sumter."

The vicinity was a woody, lonely, deserted spot. The strange and rapid changes that have actually taken place in a region, so few years since almost uninhabited and nearly unknown, seem incredible.

One day, a silent wilderness; another, a busy camp—a horrible human slaughter-house; still another, and it is the noted graveyard of America—we might say, of the world.

The so-called "Confederate States Military Prison" was a stockade made of pine logs, planted in the ground perpendicularly, so that they were about twenty feet high from the ground. There were two stockades, the first of unhewn logs, the second (being the stockade proper) of hewn timber, covering an area of 23½ acres. Inside of this was a light railing, at a distance of about twenty feet from, and running parallel with, the four sides of the square, called the " dead line. Any prisoner passing this line, by any pretence or accident, endangered his life. The space occupied by the prisoners was thus quite materially decreased.

There were two entrances; one east, near the north and south ends of the stockade, consisting of massive gates, opening into spaces about 30 feet square, on the principle of a canal lock.

On the inner stockade, at intervals of say ten rods, were sentry boxes, covered so as to protect the inmates from storm and sun. The rebel guard stationed in these boxes were so elevated as to have a perfect view of all that was taking place within.

The object of the two stockades was, that if attacked, the rebel force acting as guard might defend themselves behind the outside wall, while the prisoners should still be confined within the inner wall.

At certain angles of this outer wall small parapets were thrown up, in the shape of angular forts, in some of which artillery was placed, commanding the prison grounds, as well as the open fields surrounding the stockade.

A small stream passed through the centre of the stockade, on each side of which the land gradually ascended to a height of about forty or fifty feet, so that the camp was really upon two side hills. At the head of this stream, outside and in immediate proximity to the inner stockade, was the main cook house of the prison. It was a wooden, barn-like building, covering some immense cauldron kettles and some very large but very poor ovens. The cooking of eatables was a mere farce.

A short time before this camp was deserted, another cook house was erected, but was used only a few times.

In the early history of the prison, the hospital consisted of a space of ground about four rods square, in one corner of the stockade. A favored few were allowed to lay under some tarpaulins, stretched over poles, placed horizontally on forked stakes.

The latter part of May, the stockade becoming crowded, and the number of sick being largely on the increase, a board enclosure, covering about five acres, was put up on the south side of the stockade, and called the hospital. The worst cases were then removed to this enclosure. A small stream of water ran through the south end, and a cook house (that is, a kettle) was placed near this stream. A few tents and several pieces of canvas constituted the shelter for the sick. This hospital would accommodate about 1,000. It generally had 2,500 inmates.

The care of the sick at the Hospital was given to Federal prisoners, and there was in this camp an attempt at order and decency.

The trees were not allowed to be cut down, and the shade was one of the blessings which the sun scorched invalids longed for. The camp was laid out in squares, and the streets were policed every day.

The surgeon in charge placed his command first in four divisions, a surgeon in charge of each division; second, each division in five wards, a surgeon to each ward.

Over each ward was placed a "Yankee steward," whose duty it was to stand between the rebels and their sick comrades. Under his direction, each ward steward and half a dozen nurses gave constant attendance to the sick.

Had they been furnished proper and sufficient shelter, food and medicine, the mortality list would never have reached the marvellous number that it did.

At the northwest of the stockade a shed was built, and called the "dead house." To this all the bodies were removed both from the stockade and from the hospital, and after a description was taken of the dead, they were numbered and then removed either to the dissecting sheds, or carried in wagons, about twenty to twenty-five in each load, piled up as a farmer would load in a quantity of butchered hogs, to the "Graveyard." The Graveyard was in the most pleasant location, and one might almost say the most desirable of any of the several institutions which went to make up Andersonville. It was on an elevated spot of ground, laid out in streets and squares.

In this connection it would be proper to introduce a

letter written by Miss Clara Barton in reply to my inquiry concerning the United States Cemetery at Andersonville:

OFFICE OF CORRESPONDENCE WITH THE FRIENDS
OF THE MISSING MEN OF THE UNITED STATES
ARMY, WASHINGTON, D. C., February 2d, 1869.

DEAR SIR:

In reply to your letter I send a copy of Atwater's list of the dead of Andersonville, which contains my report of those prison grounds as I found them in July, 1865. It is as complete as I could make it, and correct, I believe, in every particular.

Upon the departure of the party accompanying me, a guard was stationed and a superintendent appointed and sustained by the Government, whose duties were to keep the place and its surroundings with as little change as possible, and I presume that with the exception of the natural decay, it remains to-day nearly as described in my report.

Although I have written much, very much, in reference to prisons and prisoners, it has been of a private nature, addressed and sent to the friends of those who had suffered and died there, and not published.

I have never published a "Book" upon prisons, as many suppose, although I have written enough upon the subject to constitute the material for a number of books; but I have always considered that the prisoners themselves were the proper persons to place the woes of their prison life before the public, and that if there was a call for anything of that nature, the privilege of meeting it, and the profits accruing therefrom of right belonged to them.

Regretting that I have not more information, I can only refer you to such authors as have written upon the subject, viz., Abbott, Spencer, Hamlin (Martyria) and others whose works are well known and easily found.

You ask for my "bill." I had hoped that all my friends, at least, thoroughly understood the basis upon which I have done my little work, and that not only no bills had ever passed *out* of my office, but that no money for services or information rendered had ever been permitted to come *into* it and remain there. I have always promptly returned every dollar and half dollar that a sometimes grateful party

A HOSPITAL TENT.

would insist upon enclosing to me. The little I have been able to do for those who suffered in our country's cause, has been done for the love of it, and right and humanity.

If any opportunity present in which I can serve you to more purpose than I have been able to do in this, please let me know, and oblige,

Yours, very truly,   CLARA BARTON.

In the report referred to in Miss Barton's letter, we find the following description of the present condition of Andersonville Graveyards:

The cemetery, around which the chief interest must gather, is distant about 300 yards from the stockade, in a northwesterly direction. The graves, placed side by side in close continuous rows, cover nine acres, divided into three unequal lots by two roads which intersect each other nearly at right angles. The fourth space is still unoccupied, except by a few graves of "Confederate" soldiers.

No human bodies were found exposed, and none were removed. The place was found in much better condition than had been anticipated, owing to the excellent measures taken by Major General Wilson, commanding at Macon, and a humane public-spirited citizen of Fort Valley, Georgia, a Mr. Griffin, who, in passing on the railroad, was informed by one of the ever-faithful negroes that the bodies were becoming exposed, and were rooted up by animals. Having verified this statement, he collected a few negroes, sank the exposed bodies, and covered them to a proper depth. He then reported the facts to General Wilson, and requested authority to take steps for protecting the grounds. That patriotic officer visited Andersonville in person, appointed Mr. Griffin temporary superintendent, and gave him such limited facilities as could be furnished in that destitute country. It was determined to enclose a square of fifty acres; and at the time of our arrival the fence was nearly one-third built, from old lumber found about the place. He had also erected a brick kiln, and was manufacturing brick for drains to conduct the water away from the graves, and protect and strengthen the soil against the action of heavy rains. We found Mr. Griffin, with a force of about twenty negroes and a few mules at work on the grounds. I have understood that that gentleman furnished the labor at his own cost, while General Wilson issued the necessary rations.

The part performed by our party was to take up and carry forward the work so well commenced. Additional force was obtained from the military commandant at Macon for completing the enclosure and erecting the head boards. It seems that the dead had been buried by Union prisoners, paroled from the stockade and hospital for that purpose. Successive trenches, capable of containing from 100 to 150 bodies each, thickly set with little posts or boards, with numbers in regular order carved upon them, told to the astonished and tear-dimmed eye the sad story of buried treasures. It was only necessary to compare the number upon each post or board with that which stands opposite the name on the register, and replace the whole with a more substantial, uniform and comely tablet, bearing not only the original number, but the name, company and regiment, and date of death of the soldier who slept beneath.

I have been repeatedly assured by prisoners that great care was taken at the time by the men to whom fell the sad task of originally marking this astonishing number of graves, to perform the work with faithfulness and accuracy. If it shall prove that the work performed by those who followed, under circumstances so much more favorable, was executed with less faithfulness and accuracy than the former, it will be a subject of much regret, but fortunately not yet beyond the possibility of correction. The number of graves marked is 12,920. The original records captured by General Wilson, furnished about 10,500 ; but as one book of the record had not been secured, over 2,000 names were supplied from a copy (of his own record) made by Mr. Atwater in the Andersonville prison, and brought by him to Annapolis on his return with the paroled prisoners.

Interspersed throughout this Death Register, were 400 numbers against which stood only the dark word " unknown." So, scattered among the thickly designated graves, stand 400 tablets, bearing only the number and the touching inscription, " Unknown Union Soldier."

Substantially, nothing was attempted beyond enclosing the grounds, identifying and marking the graves, placing some appropriate mottoes at the gates and along the spaces designed for walks, and erecting a flagstaff in the centre of the cemetery. The work was completed on the 17th of August, and the party took the route homeward by way of Chattanooga, Nashville and Cincinnati, arriving at Washington on the morning of August 24th.

CHAPTER XII.

FACTS AND THEORIES.

Thirty-two thousand men were confined in an area of twenty-three and one-half acres of land.

To sustain this statement, which, I believe from observation to be correct, I would refer to Documents Nos. 1 and 4 in the Appendix. Official statements of Inspectors must, in such cases, be uncontrovertible evidence.

No shelter or protection from heat, cold or rain was furnished to the prisoners. This assertion is made by Colonel Chandler, Dr. Jones, Dr. Roy, and a host of other Confederate officers; and I, having the evidence of my own eyesight, do endorse it, and believe that there is not a survivor of Andersonville living to-day who would contradict such assertion.

A limited number of the prisoners were the fortunate possessors of army blankets.

The erection of a tent consisted in stretching one of these blankets over a pole which had been laid horizontally in two forked stakes, driven some feet or so into the ground. These quarters furnished sleeping accommodations for from four to six men.

Others, by bribing guards and going out occasionally in the squads sent outside for fuel, obtained boughs and branches, with which they framed a shelter.

Others dug holes in the side hill, sufficiently large to cover the head and shoulders, and deemed themselves happy in the possession of such a tenement.

A greater portion of the inmates of that prison had no place to rest their sick and weary bodies in day or night, except upon the hot sand or the muddy swamp, with naught but the canopy of heaven to cover them.

The supply of water was insufficient. The water was impure, even vile. (See Documents Nos. 1, 4 and 8 in Appendix.) It was not fit for bathing purposes, to say nothing of being obliged to drink it and cook with it. My own knowledge of this fact is constituted on this : that I often went to bathe and to obtain water for cooking and drinking purposes. The stream ran through the centre of the stockade. On either side was a marshy strip of ground, extending about ten feet each way, and following the stream its entire length. This morass was the general sink of the camp. Therefore, there was but one point at which water could be obtained, which even a burning thirst could force down our throats. That was at the head of the stream, and in immediate proximity to the "dead line." " A thousand men an hour at one spring of water." Realize that fact, my reader, and you may comprehend one of our difficult undertakings. Not unfrequently would prisoners endanger their lives, by reaching over the " dead line" and plunging their cups and buckets, in hoping to obtain a little purer water, and avoid a weary waiting in the line. Some were shot there.

Below this point many bathed and washed their clothing. The lampblack soot that settled over their bodies and clothing—the dirt clinging to these from their sandy and muddy beds—the accumulation of body lice and other vermin—were all removed by washing in this stream, if removed at all. The natural result of an attempted purification of such an army of persons in so small a stream, which was at the same time receiving the drainage of the marshy sink, was, that the water became thick and sluggish with such a conglomeration of filth.

Nor was this all. I have said that the purest water, which was used for drinking, was sought for at the head of the stream, within the stockade "dead line." Outside of the stockade, and above this portion of the stream, I saw, many times, camps of the rebel guard stationed on the banks of the same creek. All their refuse floated down until it reached the cook house (which was built on this same stream, near to the stockade, and within forty feet of the point where the water was obtained by the prisoners;) there it received the additional offscourings and offal of that filthy place, and the whole accumulated mass poured under the stockade timbers into the cups of thirsty men. None but men with parched and fevered throats could have drank it.

In the latter part of their stay a few wells were dug by the soldiers themselves. As a rule, however, they belonged to a firm of speculative individuals, who laid tribute (and considering the labor incurred, having nothing but their hands, tin cups and half canteens to dig with, the charge was not unjust) on others for the use of the water from these wells. The supply would have soon been exhausted if they had permitted a general use of the wells.

The quantity of food issued to the prisoners was so meagre as to gradually induce death from starvation, and the quality was such that none but starving men could have been induced to eat it. It was repulsive even to them.

The United States soldier receives a daily ration of ¾ pound of bacon, 1¼ pound of fresh or salt beef, 18 ounces of bread and flour, or ¾ pound of hard bread, or 1¼ pound corn meal, with rice, beans, vegetables, coffee, sugar, tea, &c., in proportion.

These were also the daily rations furnished the Confederate prisoners by the United States Government. Compare them with the pitiable allowance of food at Andersonville, *i. e.*, 3 to 4 ounces of spoiled bacon, half pint of meal or a piece of meal cake, composed of water and ground corn, husks and cob, either partially baked or quite burned, its cubical dimensions being, say three inches wide, one inch thick, and four to five inches long. In addition to this, we only occasionally received a small quantity of rice and a tablespoonfull of molasses, or a few worm-eaten beans, which was often termed "red maggot soup."

In the hospital the food was the same as in the stockade, with the exception of beef soup once or twice a week; and toward the latter days a gill of wheat flour was distributed perhaps twice a week.

Rations were frequently distributed raw, and no fuel provided to cook with.

These rations, which were daily dealt out to us, and called food, did not satisfy hunger. They created hunger. The corn bread, which was the staple article, after

ISSUING RATIONS AT ANDERSONVILLE.

it was masticated and swallowed with much difficulty, only irritated the stomach and bowels and produced diarrhœa, and as more was taken into the stomach, so much the more rapidly the victim passed along the road to death. There was no nourishment in it. It might as well have been so much sawdust and water.

The bacon which was given us would have disgusted a soap-fat man.

Men were suffering and dying for want of acids and vegetables. Scorbutis arises from want of acids. It can be easily cured by a proper supply of vegetables and fruits. None were ever dealt out to us. Scorbutis was the great scourge of the camp.

A very few times some cabbages were sent to the hospital. The country was full of sweet potatoes, and yet the prisoners saw none in Andersonville until the last week or two that they were there.

The supply of fuel was irregular and entirely inadequate. It was generally obtained by a squad of the prisoners detailed each day or twice a week. Not over forty were permitted to go out at a time. And this number of men were obliged to procure from the woods, and bring in upon their backs, the daily supply of branches which constituted the fuel for the use of from 15,000 to 30,000 men. Take forty sticks of wood from your "four foot" wood pile, and so splinter it that it shall make 15,000 pieces, and you have an Andersonville "ration" of wood.

After having used every influence and means to produce sickness, no proper or adequate measures were taken to cure or even to alleviate suffering. (See Documents Nos. 2, 3, 4, 5 and 7 in Appendix.)

The hospital was overcrowded. Its accommodations were the poorest imaginable.

There were a few good tents—more that were rotten and torn. No bedding, not even straw, to lie upon. Those who owned blankets could use them for a bed. Those who did not, had the ground for a couch.

One line of tents used for those who had had surgical operations performed upon them, was furnished with board bunks. But they soon became so filthy, from want of change of bed clothing, that no person, with the slightest flesh wound, dared to locate himself there, for fear of being contaminated with gangrene, which, if once possessed, doomed a man to certain death.

The food furnished to those sick men was just that which they ought not to have had.

All persons know that careful nursing and proper diet have much to do with the recovery of an invalid. The food and shelter which were furnished at that hospital would have defeated the skill of the best physicians in the world, with every remedy named in the Pharmacopœia at his command. What, then, could be expected of half-fledged physicians (as most of the prison surgeons were), with little other than indigenous medicines?

There were a few able physicians and excellent surgeons—men with kind hearts and much sympathy, but they were powerless.

The supply of medicine was so small, that the dispensary would be unable to supply the smallest requisitions for several days in succession.

None of the stores usually found in hospitals were ever seen there, notwithstanding such things were sent from our Northern friends by the Christian Commission. (See Document No. 9 in the Appendix.)

Flour was the only luxury ever rationed out to the inmates of the hospital; and what could sick men do with flour, having no fire and no utensils to cook with?

The best evidence that can be had that these assertions are true, is the mortality that occurred at that prison. Thirteen thousand died in about eight months.

This tells the tale of hunger and thirst, of disease and suffering, of want of comfort and care, of lack of nourishment and medicine, in words of brevity, but words of terrible meaning.

Notwithstanding all this unjust and cruel treatment, still did they lengthen their list of crimes by adding cruelties under the name of punishments.

Several times they ceased to issue rations for a day and even two days. Cause—"some few of our number had dug a tunnel in order to escape," and to punish these men thousands of starving men were deprived the morsel that would barely keep the breath of life in them from day to day.

They shot men. Cause—they had reached over the dead line for water, or for a cracker that was a foot beyond the dead line. And they shot men even within the dead line. I myself attended a man in the hospital who died from the effects of a wound in the leg, made by a rebel guard shooting him while he lay about the hospital camp fire, inside the dead line. I was with him a minute after the report of the musket was heard, and he had not moved from the spot he was in when he was shot.

They chased men with dogs, and these dogs did bite and mutilate men, from the effects of which they died. Cause—They were attempting to escape.

They put prisoners in chain gangs and in stocks; they whipped them at a whipping post; they hung them up by the thumbs. Cause—these prisoners attempted to escape.

They did force prisoners to be vaccinated with poisonous virus, and but few that were vaccinated lived.

They destroyed letters to our homes and letters from our homes, uselessly, carelessly, and purposely to distress us. They destroyed or themselves used great quantities of clothing, food and delicacies that were sent to us by express from the North.

They beat and kicked sick soldiers who were too ill to keep up in line of march. And, last of all, when they had killed by inhuman treatment and cruelties, they buried our friends and comrades in an indecent manner that even barbarians could not have excelled.

Starvation, thirst, want of clothing and shelter, cruel treatment, disease, want of medicine and medical attendance, and lastly an indecent burial, are a terrible and revolting list of horrors; yet there was still another trial that to a prisoner was harder to bear than all these: that was the agony of the mind which was caused by the knowledge that our imprisonment might be a long one, and death was certain if we were held there any length of time. There was nothing to look forward to. Day followed day, and all were alike. Nothing to divert the mind, no exercise for the body. Sorrow and despair pervaded the camp. A smile was a rare thing, a real laugh almost unknown. There were hollow, forced laughs, that went back down our throats, like the resounding tones of a voice in a dark, damp cavern. They caused us rather to shudder than to feel merry.

No stories had interest, they but recalled the time when we were free. The mind was left to itself, and it would destroy itself. Depression and homesickness were the terrible forms of disease that we feared. Once under the influence of either in that terrible place, we could with greatest difficulty rise from it, so insinuating and so gradual was its approach. Thoughts of home became constant. Dreams of home and of home comforts, especially of the favorite dishes that had been prepared by the hands of a doting mother, a pet sister, or a loving wife, were of nightly and even daily occurrence. Then the victim began to talk of home, of the probabilities of an immediate exchange of prisoners, of the probable exertions that were being made for his release. He begins to believe that he will soon be exchanged. He looks at every rebel guard that approaches the gates with the expectation that they are coming to free him. He talks of nothing else, his mind cannot be directed from that one subject.

Days pass by and yet he is not sent for, still he believes and watches. He sits in the wide camp street, where his eye may rest constantly on the gate, refuses food, refuses to move from his position night or day. His mind wanders, his eye is vacant and staring, he is weak, and though in sitting posture, falls over to the ground.

"There, there they come! I said they'd come, I knew they'd come! Hold me up, I must go to meet them. Mother! father! I am glad you've come. I'm so tired, I'm so sick; take me home."

God grant that it was his good angels that came to our dying comrade, as the "king of terrors" thus made his last day one of happy delirium.

This is only one true picture out of many which I saw in Andersonville Hospital.

T. J. Hyatt, sergeant in Company K, One Hundred and Eighteenth Pennsylvania Volunteers, is the author of the following lines, which very truthfully tell the feeling that was uppermost in the hearts of us all:

THE PRISONER'S PLEA.

When our country called for men, we came from forge and store and mill,
From workshop, farm and factory, the broken ranks to fill;
We left our quiet, happy homes and ones we loved so well,
To vanquish all our Union's foes, or fall where others fell.
Now in prison drear we languish, and it is our constant cry,
Oh ye who yet can save us, will ye leave us here to die?

The voice of slander tells you that our hearts were weak with fear—
That all, or nearly all of us, were captured to the rear.
The scars upon our bodies from the musket balls and shell,
The missing legs and shattered arms, a truer tale can tell.
We have tried to do our duty in the sight of God on high;
Oh ye who yet can save us, will ye leave us here to die?

There are hearts with hope still beating, in our pleasant Northern homes,
Waiting, watching for the footsteps, that may never, never come.
In Southern prisons pining, meagre, tattered, pale and gaunt,
Growing weaker, weaker daily, from pinching cold and want,
Their brothers, sons and husbands, poor and hopeless captives lie,
Oh ye who yet can save them, will ye leave them there to die?

From out our prison gate there's a graveyard near at hand,
Where lie twelve thousand Union men, beneath the Georgia sand;
Scores on scores are laid beside them, as day succeeds to day,
And thus it will be ever, till they all shall pass away;
And the last can say while dying, with upturned and glazing eye,
Both Faith and Love are dead at home, they have left us here to die.
ANDERSONVILLE, Oct. 20, 1864.

[A singular incident has occurred in regard to this poetry. We had supposed Sergeant Hyatt was dead. We had copied the lines in his tent at Andersonville, and had afterwards been told that he died the following morning. What was our astonishment when our printers informed us that the compositor who was setting up this chapter was none other than Sergeant Hyatt, and the author of this poetry. The lines are none the less pathetic, and certainly the more romantic, from this singular circumstance.]

We had intended to follow out in this chapter a line of argument and facts as to the following points:

Was this inhuman treatment necessary?

Did not the Confederacy possess food, fuel, water, clothes and medicine, bedding, tents and lumber, and was it not immediately near Andersonville?

Did Jefferson Davis and his Cabinet know of this condition of things?

Were they not accorded in by superior officers of the Confederacy as well as the inferior officers?

Can, therefore, any blame be attached to any Federal officer by the sophistry that declining to exchange on unjust and unfair terms, was assuming the responsibility of causing this suffering?

Space will not allow us to introduce evidence on these points, but we can assert that there are quantities of proof to show that this treatment was not necessary; that there was plenty of provisions, lumber, fuel, and all other things needed to save life and health; that all this suffering was known to all prominent Confederate officials, and that it was intended to cripple the Northern army, by killing prisoners of war, or at least making them unavailable on account of chronic disease.

In reviewing this chapter, it occurs to the writer that his reader may complain that it is too general in its character to be entitled to a place in this book of records. But I must beg you to bear in mind that in this instance general experience is individual experience. The boys of the Battery suffered in the proportion of 100 to 30,000.

As a rule there was no show of weak heart or faltering will. They stood up and grappled with this monstrous horror, with the same unflinching bravery that they displayed when in battle. They were, too, mostly Christians, and death rather than dishonor was their decisive reply to overtures or taunts from rebels. They met death, if in consciousness, with calmness and even willingness. They were kindly and tenderly treated by each other. The devotion of the well to their sick comrades was notable.

If there is or ever has been on this earth a place where selfishness and self preservation even at the cost of another's life, could make an appearance, Andersonville was that place. Yet to their honor be it said, were the members of the Twenty-fourth New York Battery true and faithful to each other. Honor and generosity were triumphant over animal instinct, and death became sweet from the knowledge of devotion and sacrifice of friends.

Slander and hurtful innuendos to the contrary, we assert that to the end was there uprightness, faithfullness and affection, between all of those boys. And when that great day comes, when we shall all meet, we believe there will be joyful greetings by each to the other.

In November, the few that were left of the Battery were made to feel that there was still reason for clinging to this wretched mode of existence, since they were informed by good authority that exchanges were actually taking place. Finally the writer, with others, was placed on the cars and started towards Savannah, and told that at Savannah we would meet the Federal exchange fleet. A two days' trip, which was endured by even the sick without murmur, brought us to Millen, and the reader may imagine the terrible reaction of spirit and hopes, when we were ordered to march into the new prison stockade. It was a paradise in comparison with Andersonville, but only another dreary prison in comparison with the country which we had supposed we were bound for—our own country.

However, our stay, to our glad surprise, was of short duration. In a few days we were again called out and taken toward Savannah. Having here signed a parole of honor, we were taken in steamers to Venus Point and delivered over to the United States Exchange Agent, General Mulford.

Our joy knew no bounds. Threats of taking them back unless they were less demonstrative could hardly check the shouts of the captives as they again beheld their long missed but dearly loved Stars and Stripes. It was a proud and happy day. Reflections were sad, but anticipations were joyous. We had sad tales to tell, but we had dear friends to meet.

It was a cold, dreary winter day when we entered Perry. Familiar faces crowded about us. Anxious inquiries overpowered the warm welcome, and we felt that to be the bearer of such tidings was indeed an unenviable lot.

The statistics presented in this book will tell the story far better than any description given by a single witness. No reader of this volume can deny that on the part of these lost comrades there was a brave sacrifice to loyalty. And I would, with these last words, still beseech you to show them honor and to do them justice. Let us raise a monument to their memory.

A NIGHT IN ANDERSONVILLE.

RECORDS

OF VOLUNTEERS FROM PERRY AND VICINITY, WHO ENLISTED IN THE FEDERAL ARMY DURING THE WAR, IN OTHER ORGANIZATIONS THAN THE TWENTY-FOURTH NEW YORK BATTERY.

Under the suggestions and persuasions of friends interested in the monument enterprise, the writer presents, with some reluctance, the following incomplete and necessarily brief records. The same opportunities and the same documents for information regarding these men, as were in our possession and gave us accurate evidence regarding the members of the Battery, are not obtainable.

We know these sketches are not full, but, so far as they go, we believe them to be correct.

That these men are entitled to equal honor with our comrades in the Battery, we promptly admit, and we therefore feel that, so far as our knowledge and such information as is in our power to obtain, shall, through our pen, do them honor, it shall be done.

We appeal earnestly in behalf of those brave comrades of the war whose record is, " died while in the service," and who lost their lives for the sake of their country. Let us all, with one interest and purpose, do justice to all who left their homes and found a grave while in such noble service.

We find the following statements concerning the different organizations represented in these personal records, in the Adjutant-General's Report of the State of New York:

FIRST NEW YORK MOUNTED RIFLES.

This regiment was organized at New York City, to serve three years. The companies of which it was composed were raised in the State at large. It was mustered into the service of the United States from August 31, 1861, to September 9, 1862. The original members, except veterans, were mustered out on expiration of term of service. The regiment, composed of veterans and recruits, was consolidated with the Third Regiment New York Cavalry, July 21, 1865; the consolidated force being known as the Fourth Provisional New York Cavalry.

SECOND NEW YORK MOUNTED RIFLES.

This regiment was organized at Buffalo, N. Y., to serve three years. The companies of which it was composed were raised principally in the counties of Erie, Niagara, Wyoming, Orleans, Alleghany and Wayne. It was mustered into the service of the United States from October, 1863, to February, 1864. Mustered out of service, August 10, 1865, in accordance with orders from the War Department.

Battles—Coal Harbor, Petersburg, Bethesda Church, Weldon Railroad, Pegram's Farm, Hatcher's Run, Poplar Spring Church.

THIRD NEW YORK CAVALRY.

This regiment was organized at New York City, to serve for three years. The companies of which it was composed were raised principally in the counties of Albany, Schoharie, Chemung, Delaware, Oneida, Onondaga and Orleans. It was mustered into the service of the United States from July 17 to August 27, 1861. On the expiration of its term of service the original members, except veterans, were mustered out, and the regiment, composed of veterans and recruits, retained in service. It was consolidated with the First Mounted Rifles, July 21, 1865. The consolidated force was designated the "Fourth Provisional Cavalry."

Battles—Burn's Church, Young's Cross Roads, Williamston, Kinston, Whitehall, Goldsborough, Ball's Bluff, Weldon Railroad, Edward's Ferry, Stony Creek, Petersburg, Malvern Hill, Newmarket, Johnson's House.

NINTH NEW YORK CAVALRY.

This regiment was organized at Albany, N. Y., to serve three years. The companies of which it was composed were raised in the counties of Chautauqua, Cattaraugus, Wyoming, Rensselaer, Washington, St. Lawrence and Clinton. It was mustered into the service of the United States from September 9 to November 19, 1861. On the expiration of its term of service the original members, except veterans, were mustered out. The Fourth Regiment New York Cavalry was transferred to this regiment as Companies B, E and L, and the organization, composed of veterans and recruits, retained in service until July 17, 1865, when it was mustered out of service in accordance with orders from the War Department.

Battles—Cedar Mountain, Brandy Station, Aldie, Upperville, Gainesville, Bull Run, Chantilly, Antietam, Gettysburg, Kelly's Ford, Rappahannock Station, Sulphur Springs, Opequan, Wilderness, Coal Harbor, Mechanicsville, Deep Bottom, Winchester, Fisher's Hill, Cedar Creek, Petersburg, Richmond.

THIRD NEW YORK ARTILLERY.

This regiment (originally Nineteenth Regiment Infantry) was raised at Auburn, N. Y., and was mustered into the service of the United States, May 22, 1861, to serve two years. It was reorganized as the Third Artillery, January 31, 1862. On the expiration of its term of service the original members were mustered out, and the organization, composed of veterans and recruits, retained in service. The Twenty-fourth Battery was assigned to this regiment as Company " L," March 5, 1865. The regiment was mustered out by batteries from June 22 to July 29, 1865, in accordance with orders from the War Department.

EIGHTH NEW YORK HEAVY ARTILLERY.

This regiment (originally the One Hundred and Twenty-ninth Regiment New York Volunteer Infantry) was organized at Lockport, N. Y., to serve three years, and mustered into the service of the United States as such, August 22, 1862. It was changed to an artillery regiment in February, 1863. Two additional companies were organized for this regiment in January, 1864. The whole organization was raised in the counties of Niagara, Orleans and Genesee (29th Senate District). Companies G, H, I and K were transferred to the Fourth New York Artillery, June 4, 1865. Companies L and M were transferred to the Tenth New York Volunteer Infantry, and the remaining six companies mustered out June 5, 1866, in accordance with orders from the War Department.

Battles—Spottsylvania, Tolopotomoy, Coal Harbor, North Anna, Petersburg, Strawberry Plains, Deep Bottom, Ream's Station, Boydton Road.

SEVENTEENTH NEW YORK VOLUNTEER INFANTRY.

This regiment was organized at New York City, to serve two years, The companies of which it was composed were raised in the counties of New York, Westchester, Rockland, Wayne, Wyoming and Chenango. It was mustered into the service of the United States, May 20 to 24, 1861. Mustered out June 2, 1863, by reason of expiration of term of service. The recruits enlisted for three years were transferred to the Twelfth Regiment New York Volunteers.

Battles—Hanover Court House, Groveton, Fredericksburg.

TWENTY-FIRST NEW YORK VOLUNTEER INFANTRY.

This regiment was organized at Elmira, N. Y., to serve two years. The several companies composing it were raised in Buffalo, N. Y. It was mustered into the service of the United States, May 20, 1861. Mustered out by reason of expiration of term of service, May 18, 1863.

Battles—Second Bull Run, South Mountain, Antietam, Fredericksburg.

TWENTY-SEVENTH NEW YORK VOLUNTEER INFANTRY.

This regiment was organized at Elmira, N. Y., to serve for two years. The several companies of which it was composed were raised in the counties of Alleghany, Broome, Livingston, Monroe, Orleans, Wayne and Westchester. It was mustered into service, May 21, 1861. Mustered out by reason of expiration of term of service, May 21, 1863.

Battles—Bull Run, Gaines' Mills, Seven Days' Battle, Second Bull Run, South Mountain, Antietam, Fredericksburg, Marye's Heights.

THIRTY-FIRST NEW YORK VOLUNTEER INFANTRY.

This regiment was raised and organized in New York City. It was mustered into the service of the United States, May 24 to June 14, 1861. Mustered out by reason of expiration of term of service, June 4, 1863.

Battles—Bull Run, West Point, Gaines' Mills, Charles City Cross Roads, Malvern Hill, Crampton Gap, Antietam, Fredericksburg, Marye's Heights.

THIRTY-THIRD NEW YORK VOLUNTEER INFANTRY.

This regiment was organized at Elmira, N. Y., to serve for two years. The companies of which it was composed were raised in the counties of Livingston, Ontario, Seneca, Wayne and Yates. It was mustered into the service of the United States, May 22, 1861. Mustered out by reason of expiration of term of service, June 2, 1863.

Battles—Lee's Mills, Williamsburg, Mechanicsville, Gaines' Mills, Savage Station, Crampton Gap, Antietam, Fredericksburg, Marye's Heights, Salem Heights.

THIRTY-SIXTH NEW YORK VOLUNTEER INFANTRY.

This regiment was organized in New York City, to serve two years. The companies comprising it were raised in the counties of New York and Erie. It was mustered into the service of the United States, June 17 to July 4, 1861. Mustered out, July 5, 1863, on expiration of term of service.

Battles—Seven Pines, Malvern Hill, Marye's Heights, Salem Heights.

EIGHTY-NINTH NEW YORK VOLUNTEER INFANTRY.

This regiment was organized at Elmira, N. Y., to serve three years. The companies of which it was composed were raised in the counties of Broome, Chenango, Delaware, Livingston, Monroe and Schuyler. It was mustered into the service of the United States, December 6, 1861. On the expiration of its term of service the original members (except veterans) were mustered out, and the regiment, composed of veterans and recruits, retained in service until August 3, 1865, when it was mustered out in accordance with orders from the War Department.

Battles—Suffolk, Camden, South Mountain, Antietam, Fredericksburg.

ONE HUNDRED AND THIRTIETH NEW YORK VOLUNTEER INFANTRY, OR NINETEENTH NEW YORK CAVALRY, OR FIRST NEW YORK DRAGOONS.

This regiment was organized at Portage, N. Y., to serve three years. The companies of which it was composed were raised in the counties of Wyoming, Livingston and Alleghany (30th Senate District). It was mustered into the service of the United States, September 3, 1862. Changed to Nineteenth Cavalry (First Dragoons) August 11, 1863.

ONE HUNDRED AND THIRTY-SIXTH NEW YORK VOLUNTEER INFANTRY.

This regiment was organized at Portage, N. Y., to serve three years. The companies of which it was composed were raised in the counties of Alleghany, Livingston and Wyoming (30th Senate District). It was mustered into the service of the United States, September 26, 1862. Mustered out, June 13, 1865, in accordance with orders from the War Department.

Battles—Chancellorsville, Gettysburg, Lookout Mountain, Chattanooga, Missionary Ridge, Knoxville, Buzzards' Roost Gap, Resaca, Cassville, Dallas, Gilgal Church, Kulp's Farm, Kenesaw Mountain, Peach Tree Creek, Turner's Ferry, Atlanta, Milledgeville, Savannah, Charleston, Averysburg, Bentonville, Goldsboro', Raleigh.

List of Volunteers from Perry and vicinity, who enlisted in the Federal Army during the Rebellion, in other organizations than the Twenty-fourth N. Y. Battery.

1 Ayers, Oscar.
2 Andrews, Rob't F.
3 Axtell, Abner.
4 Audrus, Merritt.

5 Beardsley, Edwin H.
6 Buttre, Chauncey.
7 Bishop, I. G.
8 Babcock, Orso.
9 Boughton, Arthur.
10 Boughton, Myron.
11 Beardsley, Alton.
12 Bullard, Rob't F.
13 Booth, Harrison.
14 Burden, Albert.
15 Burden, Adelbert.

16 Calkins, Melatiah.
17 Childs, Reuben.
18 Cady, Geo. E.
19 Chapin, Abner B.
20 Chapin, Willard J.
21 Cronkhite, Joel.
22 Crocker, Emory F.
23 Crocker, Chas. H.
24 Childs, Lucius.

25 Dunn, John.

26 French, Myron.
27 Fitch, William.
28 Flint, J. Nelson.
29 Farden, Francis.
30 Frayer, Andrew.

31 Griffith, Willis.
32 Gardner, Avery.
33 Grigg, Wm. Jr.

34 Hollenbeck, Wallace.
35 Hill, Wm.
36 Hunt, Chas. H.
37 Higgins, Frank.
38 Hunt, Geo. S.
39 Hershey, Andrew.
40 Hildren, James.

41 Jeffres, Capt. C.

42 Keeton, Jno
43 Keeney, Anson.

44 Lacy, James.

45 Matteson, Henry.
46 Mohannah, Wm.
47 Mohannah, Barton.

48 Noonen, Wm.

49 Post, Thos. E.
50 Pettibone, Levi.
51 Pettes, F. W.
52 Post, Lucius H.
53 Post, J. Mort.

54 Robinson, Jno. P.
55 Robinson, Zeb. C.
56 Robinson, Adolphus

57 Sweet, Chas.
58 Summy, David.
59 Simmons, Alpheus.
60 Simmons, Jas. B. B.
61 Simmons, Phineas A.
62 Senter, Lucius.
63 Salisbury, M. S.
64 Summy, Mort.
65 Sherman, Seymour

66 Tallman, Walter.
67 Tallman, Benj. H.

68 Westbrook, Jno.
69 Westbrook, Geo.
70 Westbrook, Nehemiah.
71 Wilson, Jno. A.
72 Williamson, Jas.
73 Westlake, Chas. G.

74 Young, Harry (col'd)

PERSONAL SKETCHES.

1. AYERS, OSCAR.—Co. H, Seventeenth New York. No further information obtained.

2. ANDREWS, ROB'T F.—At the time of the breaking out of the rebellion he was in the Western States. He enlisted in a Western light artillery battery and did good service. He remained in the army for the full term of his enlistment, then returned to Perry, and remained for some time. Is now living in Chicago.

3. AXTELL, ABNER.—Enlisted in New York City, November 15th, 1861, in the Fifth Pennsylvania Cavalry. Died, while in the service, at Georgetown, D. C., April 22d, 1862.

4. ANDRUS, MERRITT.—Enlisted in New York City, in the Fourth U. S. Artillery. Mustered out at Nashville, Tenn. Married Miss Josephine Lacy, and is now settled in Perry.

5. BEARDSLEY, EDWIN H.—Enlisted at Warsaw, in the Seventeenth New York Volunteer Infantry. Was commissioned second lieutenant, August 30, 1862. Was

promoted to first lieutenant, October 22, 1862. Served his full time. Is married, and has settled in the West.

6. BUTTRE, CHAUNCEY.—Enlisted in the One Hundred and Thirtieth New York Infantry, at Perry, N. Y., August, 1862. Was mustered out in Rochester, July 6, 1865. Is married, and settled in the Western States.

7. BISHOP, I. G.—Enlisted in the First New York Mounted Rifles, at Perry, August 18, 1862, but on account of physical inability was rejected.

8. BABCOCK, ORSO.—Enlisted in the One Hundred and Thirtieth New York Infantry. Was mustered out from the hospital, and now lives in Moscow, N. Y.

9. BOUGHTON, ARTHUR.—Enlisted in the Eighty-ninth New York Volunteers at Perry, December 16th, 1861. Died, while in service, at Roanoke Island, 1862.

10. BOUGHTON, MYRON.—We cut the following obituary from the Wyoming *Times* of November 7, 1862:

OBITUARY.—Killed at the battle of Perryville, Ky., October 8th, Myron Boughton, son of Deacon J. S. Boughton, of this village, aged 39 years.

This is the second time that Mr. Boughton has been called to mourn the loss of a son (within a few months) both of whom gave themselves as a sacrifice to their country. Mr. B. received a letter from Myron a short time since saying that he had enlisted, and about a week after another letter was received, written by stranger hands, announcing the death of his son as above. Deceased was a member of the Twenty-first Wisconsin regiment, and was mustered into service the 6th of September. He leaves a wife and three children.

11. BEARDSLEY, ALTON.—Enlisted in Company K, Seventeenth New York Infantry Volunteers, at Warsaw, May 20, 1861. Was mustered out at New York City, June 2, 1863. Married January 16, 1864, and now lives in Perry.

12. BULLARD, ROBERT F.—Enlisted in the One Hundred and Thirty-sixth New York Volunteer Infantry at Cornsus, August 28th, 1862. Was mustered out at Madison, Wis., September 5th, 1864. Was wounded at Lookout Mountain, November 23d, 1863. Promoted to Second Lieutenant, January 16th, 1863. Married, September 4th, 1862, to Miss S. E. Rosenkrans. At present living in Perry.

13. BOOTH, HARRISON.—Not able to trace particulars.

14. BURDEN, ALBERT.—Enlisted in the Second New York Mounted Rifles, at Castile, December, 1863. Mustered out at Buffalo, July, 1865.

15. BURDEN, ADELBERT.—Enlisted in the Second New York Rifles, at Perry, January, 1864. Mustered out at Buffalo, July, 1865. Is now living in Kansas.

16. CALKINS, MELATIAH—Not able to trace particulars.

17. CHILDS, REUBEN.—Enlisted in the Thirty-third New York Volunteers, April 22d, 1861, at Geneseo. He died of typhoid fever, October 27th, 1862. He died while on a furlough, at his own home.

18. CADY, GEORGE E.—Enlisted in the Twenty-seventh New York Volunteer Infantry, May, 11th, 1861, at Mount Morris. Served out his full time. Wounded at Gaines Hill.

19. CHAPIN, ABNER B.—Entered the army in the Quartermaster's Department. Was with Sherman on his "March to the Sea." Saw some of the prison pens described in this book. Is now teller in the Franklin Bank, Cincinnati, O.

20. CHAPIN, WILLARD J.—Was contract Surgeon, on duty at the hospital at Louisville, Chattanooga and Atlanta, and also with the Ambulance Train of Sherman's Army. Is now married and settled in Perry.

21. CRONKHITE, JOEL.—Enlisted in the One Hundred and Thirtieth New York Volunteers, at Perry, August 11th, 1862. Mustered out at Rochester, July 5th, 1865. Is now residing in Perry.

22. CROCKER, EMORY F.—Enlisted in the First New York Dragoons, at Perry, March 18th, 1864. Mustered out, July 1st, 1864, at Alexandria, Va. Is now living in Perry.

23. CROCKER, CHARLES H.—Enlisted in the First New York Dragoons, at Perry, August, 1862. Mustered out at Rochester, N. Y., July 6th, 1865. Now lives in Warsaw, N. Y.

24. CHILDS, LUCIUS.—No trace of his enlistment.

25. DUNN, JOHN.—Enlisted in the Eighty-ninth New York Volunteers, Perry, September, 1861. Mustered out at Washington, D. C. Served out his time in full, and now lives in Perry.

26. FRENCH, MYRON.—Enlisted in the One Hundred and Thirty - sixth New York Infantry, at Portage, August, 1862. He is reported as having died at Stafford C. H., Va., April, 1863. In our school days, French was an excellent scholar, and a favorite among his fellows. We had lost track of him after he left the academy, and it was with sorrow that in searching for these records, we, for the first time, learned that he too was sacrificed on this unholy altar.

27. FITCH, WILLIAM.—No trace of his enlistment.

28. FLINT, J. NELSON, who spent his boyhood among us, will be remembered by many of our citizens, and especially by those who passed their student life at Perry Academy during its palmiest days. He writes us, that he still regards Perry as his adopted home, and recalls as the happiest period of his life, his school day associations with Keeney, Hershey, Yeckley, Barnum,

Erricson, Deverell, Chapin, Moore, Wolf, Riddell, and a score of others. He finished his academic studies at Perry Academy in 1857. Graduated at Yale College in 1861. Enlisted in the One Hundred and Thirtieth New York Volunteers (afterwards First New York Dragoons), at Portage, N. Y., August, 1862. Shared the successes and disasters of the Army of the Potomac, until the termination of the Rebellion at Appomattox Court House, April 9, 1865. Was promoted successively to sergeant, sergeant-major, second lieutenant and first lieutenant. During the campaign of 1864, was detailed as aid-de-camp to Major-General Sheridan. Was brevetted both by the Governor of New York and by the President, for gallantry in the field. A short time ago, we had the pleasure of shaking hands with him in Elko, a mining town of Eastern Nevada, where he now is.

29. FARDEN, FRANCIS.—Enlisted in the Eighth New York Heavy Artillery, at Castile, December 28, 1863. Mustered out at Washington, D. C., September 26, 1865. Now living in Iowa.

30. FRAYER, ANDREW.—Enlisted in the Eighth New York Heavy Artillery at Castile, December 28, 1863. Mustered out at Washington, D. C., September 26, 1865. Now living in Iowa.

31. GRIFFITH, WILLIS.—Enlisted at Mt. Morris, N. Y., in the Twenty-seventh New York Volunteer Infantry, May 11, 1861. Died while in service, December 24, 1862.

32. GARDNER, AVERY.—Enlisted at Perry, in the Eighty-ninth New York Volunteer Infantry, September, 1861. Mustered out at Washington, D. C. Was wounded and re-enlisted. Now lives at Perry.

33. GRIGG, WM., JR.—Enlisted at Castile, N. Y., in the Eighth New York Volunteers, December 28, 1863. Mustered out at Washington, D. C., September 26. Now resides in Perry.

34. HOLLENBECK, WALLACE.—Enlisted at Perry, 1861, in the Ninth New York Cavalry. Served his full time, and we are informed, since his return to his home, died.

35. HILL, WILLIAM.—Enlisted at Perry, N. Y., in the Eighty-ninth New York Volunteer Infantry. Was discharged on account of being disabled by a wound received in battle. He was married to Miss Kate R. Keeney, of Perry, and is now living in Kansas.

36. HUNT, CHARLES H.—Enlisted at Mt. Morris, in the Twenty-seventh New York Volunteer Infantry, May 11, 1861. Was taken prisoner July 21, 1861. He was soon after exchanged, and returned to Perry, and died at his father's house, July 3, 1862. He was not a strong man, and doubtless his imprisonment hastened his death.

37. HIGGINS, FRANK—Enlisted in the First New York Mounted Rifles. We have no further trace of him.

38. HUNT, GEORGE S.—We copy the following from the correspondence of the Wyoming *Times*, of August 15, 1862:

I regret the painful duty of announcing the death of George S. Hunt, of Perry, who died at Mill Creek Hospital, near Fortress Monroe, on Friday, July 25th, 1862. In the spring of 1861, with many of the Perry boys, George accompanied your correspondent to Camp Scott, Staten Island, and following the fortunes of Captain Bennett's company, he returned to Elmira, where he re-enlisted in deeper earnest, and a few weeks found him asserting the authority of his flag on the bloody and unfortunate battle-field of Bull Run. George never possessed an invulnerable constitution, and the exposure incident to a soldier's life, frequently manifested itself in severe attacks of headache, on which occasions he mourned the absence of friends, particularly his mother, to whom he was deeply attached. During the week of battles, he regretted nothing more than his inability to join us in the dangers of the field, and till the end he insisted in devoting all his feeble efforts to the care of our wounded. But the malaria of the Chickahominy was in his system, and he failed under the exposure and excitement of the seven days march. Harrison's Landing offered little accommodation to the thousands of sick and wounded soldiers, but through the kindness of Captain Hall, George was removed to more comfortable quarters. Still desiring to be with his company, he soon returned to camp, where he enjoyed the hospitality of Westbrook's tent till hospital tents were erected, to which he was immediately removed.

But it was soon evident that he would not recover here, and he was removed to receive better care. His case was hopeless and he died.

His comrades here feel genuine sorrow at his death, and while they mourn the loss of a brave and faithful soldier, they have great consolation in knowing that he who braved death at the cannon's mouth, did not fear to cross the threshhold of the tomb.

39. HERSHEY, ANDREW.—He joined the East Gulf Squadron as Assistant Surgeon, 10th of July, 1863.

Died of heart disease, at Key West, Florida, February 6, 1864. Was promoted to Surgeon. As he was one of our warmest-hearted and jolliest companions in school, he had many friends, and no enemies, during our school days. His popularity seemed to follow him as he went into the world. Of strangers he soon made warm friends, and the duties which his profession demanded from him, we have no doubt, performed with a faithfulness and cheerfulness that were welcome to the sick and suffering.

40. HILDREN, JAMES.—Enlisted in the Eighth New York Heavy Artillery, at Castile, December 28, 1863. Mustered out September 26, 1865, at Washington, D. C. Now living in Perry.

41. JEFFRES, C., CAPT.—Enlisted in the Thirty-sixth New York Volunteers, at Perry, August, 1862. Mustered out at Washington, D. C., February, 1863. Is now living at Okolona, Wisconsin.

42. KEETON, JOHN.—Enlisted in the First New York Mounted Rifles, at Castile, August 13th, 1862. Mustered out at Richmond, Va., May 19, 1865. Is now living at Castile.

43. KEENEY, ANSON.—Enlisted in the Eighty-ninth New York Volunteers, at Mt. Morris, September, 1861. Mustered out at Washington, D. C., and is now living in Perry.

44. LACY, JAMES.—Enlisted in the First New York Mounted Rifles, at Perry, August, 1862, and died while in the service. Lacy was still another of the old Perry Academy school mates that was swallowed up in the maelstrom of enthusiastic patriotism, and before we hardly knew that he had gone to the war, we heard the sad tidings of his death.

45. MATTESON, HENRY.—Enlisted in the Eighth New York Heavy Artillery, at Castile, December 28, 1863. Mustered out at Washington, D. C., September 26, 1865. Is now living in Perry.

46. MOHANNAH, WILLIAM.—Enlisted in the Thirty-first New York Volunteers, at Perry, January 15, 1863. Mustered out at Hartford, Conn., January, 1865. Now living at Castile, N. Y.

47. MOHANNAH, BARTON.—Enlisted in the Thirty-first New York Volunteers, at Perry, January 15, 1863. Died while in the service.

48. NOONEN, WILLIAM.—Enlisted in the One Hundred and Thirty-sixth New York Volunteers, at Perry, N. Y., August, 1862. Was wounded at Gettysburg, July 3, 1863. Mustered out from hospital, and is now living at Perry.

49. POST, THOMAS E.—Enlisted at Auburn, N. Y., in the Third New York Artillery, at Auburn, N. Y., in the early part of the war. Was stationed in New Berne at the

time the Twenty-fourth New York Battery was there, and we thus renewed our boyhood acquaintance. Served his full time, and was mustered out at New Berne, N. C. Is now living in Buffalo.

50. PETTIBONE, LEVI.—Enlisted in the Eighty-ninth New York Volunteers, at Perry, October, 1861. Died at Roanoke, N. C., 1862, while in service.

51. PETTES, FRED. W.—Enlisted in the First New York Cavalry, July, 1861. Was transferred to the Fiftieth New York Engineers. Commissioned as second lieutenant in June, 1862. Was promoted to first lieutenant, and afterwards to captain. Is now living at Warsaw.

52. POST, LUCIUS H.—Enlisted at Warsaw, first in the Seventy-fourth New York Militia (see Salisbury's Record). Soon after enlisted from Warsaw in the Seventeenth New York. Was wounded at one of the battles in Va. Promoted to sergeant, and afterwards received a commission as lieutenant. His army correspondence to the country newspapers was quite interesting. Was married to Miss Morris, of Warsaw, and is now publishing a popular newspaper in Dekalb, Ill.

53. POST, J. MORT.—Enlisted at Rochester, July 16, 1861, in the Third New York Cavalry. Mustered out at Suffolk, Va., July 12, 1865. Re-enlisted as a veteran at Newport News, Va., December 16, 1863. Promoted from sergeant to second lieutenant, June, 1864. From

second lieutenant to first lieutenant and adjutant, July 6, 1864. From first lieutenant to captain, January 8, 1865. Married to Miss Minerva Morris, of Warsaw, January 26, 1865. Now living at Independence, Iowa.

54. ROBINSON, JOHN P.—Enlisted at Portage, August 7, 1862, in the One Hundred and Thirtieth New York Volunteer Infantry. Was mustered in as captain. Became an exceedingly popular officer. Was promoted to major in 1865, and since the war has received a colonel's commission. Was mustered out July 17, 1865, at Rochester, N. Y. In November, 1867, was elected County Clerk of Wyoming County, polling a large vote on his war record. Was married to Miss Laura Bristol, of Warsaw, and is at present living at Warsaw, N. Y.

55. ROBINSON, ZEB. C.—Enlisted in the Third New York Cavalry, at Rochester. Zeb. Robinson and Mort. Post were then attached to a company of the Third New York Cavalry, which was frequently stationed at the same place that the Twenty-fourth New York was stationed. We thus kept up quite a social acquaintance until the Battery was ordered to Plymouth. We missed their companionship. Robinson was promoted to a lieutenantcy. Served his full time in active service. The last we heard of him was, that he had just entered into a contract for life with Miss Scovill, of Rochester, N. Y.

56. ROBINSON, ADOLPHUS.—Enlisted in the One Hundred and Thirtieth New York Volunteers, at Perry,

August 11, 1862. Mustered out at Rochester, N. Y., July 15, 1865. Is now living at Perry, N. Y.

57. SWEET, CHARLES.—Cannot give particulars; but we are informed that he enlisted as a musician. On his return married Miss Julia Andrus. Cannot give his present address.

58. SUMMY, DAVID.—Enlisted in the Twenty-seventh New York Volunteer Infantry, at Mount Morris, N. Y., May 11, 1861. Served out his full time.

59. SIMMONS, ALPHEUS.—Enlisted in the Eighty-ninth New York Volunteers, at Perry, September 8, 1861. Mustered out at Washington, November 1, 1862. Was promoted to second lieutenant. Is now living in Perry.

60. SIMMONS, JAMES B. B.—Enlisted in the First New York Dragoons, at Perry, 1863. Died while in service, at Perry, September 1, 1864.

61. SIMMONS, PHINEAS A.—Enlisted in the One Hundred and Thirtieth New York Volunteer Infantry, at Perry, 1862. Died at Suffolk, Va., while in the service, October, 1862.

62. SENTER, LUCIUS.—Enlisted in the Eighty-ninth New York Volunteers, at Perry, September, 1861. Died at Roanoke Island, N. C., in 1862, while in service.

63. SALISBURY, M. S.—His first enlistment was April 25, 1861, in Buffalo, in Company C of the Seventy-fourth New York Militia. On account of no more militia regi-

ments being accepted by the Government, he immediately enlisted and was mustered in Company C of the Twenty-first New York Volunteers, May 7, 1861. Mustered out at Buffalo, N. Y., May 10, 1863.

Salisbury was the first man that enlisted from the town of Perry. He was obliged to go to Buffalo to enlist, since at that time there was no opportunity for enlisting at Perry. We deem such first enlistment quite an honor; and as we have as yet been unable to find any one who ranks him in date of enlistment, we cheerfully accord to him that honor.

Ed. Beardsley and Lucius Post enlisted about the same time from Warsaw.

64. SUMMY, MORT.—No positive information obtained. Was told that he enlisted as a musician.

65. SHERMAN, SEYMOUR.—Enlisted in the Thirtieth New York Volunteers, at Perry, January 15, 1863. Died at hospital a short time after his enlistment.

66. TALLMAN, WALTER.—Enlisted in the Eighth New York Heavy Artillery, at Castile, November, 1863. Mustered out at Rochester, August, 1865, and is now living in Castile.

67. TALLMAN, BENJ. H.—We are informed by Captain C. E. Martin, of Mount Morris, that Tallman enlisted in the Twenty-seventh New York Volunteer Infantry, May 11, 1861. Served his time out, and then enlisted in the Twenty-fourth N. Y. Cavalry.

68. Westbrook, John.—Enlisted at Geneseo, N. Y., in the One Hundred and Fourth New York Volunteer Infantry, in 1861. Was wounded while in battle. Mustered out at Smoketown. Is now living in Perry.

69. Westbrook, George.—Enlisted at Geneseo, in the One Hundred and Fourth New York Volunteer Infantry.

70. Westbrook, Nehemiah.—Enlisted at Lima, in the Twenty-seventh New York Volunteer Infantry, 1861. Mustered out at Elmira.

71. Wilson, John A.—Enlisted in the One Hundred and Thirtieth New York Volunteers. Was discharged soon after his enlistment, and is now living in Cold Water, Mich.

72. Williamson, James.—Enlisted in the Eighth New York Heavy Artillery, at Perry, December 28, 1863. Was mustered out at Rochester, N. Y., September, 1865, and is now living in Perry.

73. Westlake, Charles G.—Enlisted in the One Hundred and Thirtieth N. Y. Volunteers, at Perry, August 11, 1862. Mustered out at Elmira, June, 1865, and is now living at Perry.

74. Young, Harry (Colored).—Enlisted in the Thirty-first New York Volunteers, at Perry, January 15, 1863. Mustered out at Hartford, Conn., January, 1865, and is now living in Perry.

APPENDIX.

DOCUMENT No. 1.

LIEUTENANT COLONEL D. T. CHANDLER'S REPORT OF HIS INSPECTION OF THE ANDERSONVILLE PRISON.

ANDERSON, August 5, 1864.

COLONEL: Having, in obedience to instructions of the 25th ultimo, carefully inspected the prison for Federal prisoners of war and post at this place, I respectfully submit the following report:

The Federal prisoners of war are confined within a stockade fifteen feet high, of roughly hewn pine logs, about eight inches in diameter, inserted five feet into the ground, enclosing, including the recent extension, an area of 540 by 260 yards. A railing around the inside of the stockade, and about twenty feet from it, constitutes the "dead line," beyond which the prisoners are not allowed to pass, and about three and a quarter acres near the centre of the enclosure are so marshy as to be at present unfit for occupation—reducing the available present area to about twenty-three and a half acres, which gives somewhat less than six square feet to each prisoner. Even this is being constantly reduced by the additions to their number. A small stream passing from west to east through the enclosure, at about 150 yards from its southern limit, furnishes the only water for washing accessible to the prisoners. Some regiments of the guard, the bakery and cook house being placed on the rising grounds bordering the stream before it enters the prison, renders the water nearly unfit for use before it reaches the prisoners. This is now being remedied in part by the removal of the cook house. Under the pressure of their necessities the prisoners have dug numerous wells within the enclosure, from which they obtain an ample supply of water to drink, of good quality. Excepting the edges of this stream, the soil is sandy and easily drained, but from thirty to fifty yards on each side of it the ground is a muddy marsh, totally unfit for occupation, and having been constantly used as a sink since the prison was first established, it is now in a shocking condition, and cannot fail to breed pestilence. An effort is being made by Captain Wirz, commanding the prison, to fill up the marsh and construct a sluice, the upper end to be used for bathing, &c., and the lower end as a sink, but the difficulty of procuring lumber and tools very much retards the work, and threatens soon to stop it. No shelter whatever nor material for constructing any have been provided by the prison authorities, and the ground being entirely bare of trees, none is within reach of the prisoners, nor has it been possible, from the overcrowded state of the enclosure, to arrange the camp with any system. Each man has been permitted to protect himself as best he can, stretching his blanket, or whatever he may have, above him on such sticks as he can procure, thatches of pine or whatever his ingenuity may suggest and his cleverness supply. Of other shelter there is and has been none. The whole number of prisoners is divided into messes of 270 and subdivisions of 90 men, each under a sergeant of their own number and selection, and but one Confederate States officer, Captain Wirz, is assigned to the supervision and control of the whole. In consequence of this fact, and the absence of all regularity in the prison grounds, and there being no barracks or tents, there are and can be no regulations established for the police consideration of the health, comfort and sanitary condition of those within the enclosure, and none are practicable under existing circumstances. In evidence of their condition, I would cite the facts that numbers have been murdered by their comrades, and that recently, in their desperate efforts to provide for their own safety, a court organized among themselves by authority of General Winder, commanding the post, granted on their own application, has tried a large number of their fellow prisoners, and sentenced six to be hung, which sentence was duly executed by themselves within the stockade, with the sanction of the post commander. His order in the case has been forwarded by him to the War Department. There is no medical attendance provided within the stockade. Small quantities of medicines are placed in the hands of certain prisoners of each squad or division, and the sick are directed to be brought out by sergeants of squads daily at "sick call," to the medical officers who attend at the gate. The crowd at these times is so great that only the strongest can get access to the doctors, the weaker ones being unable to force their way through the press; and the hospital accommodations are so limited, that though the beds (so called) have all or nearly all two occupants each, large numbers who would otherwise be received are necessarily sent back to the stockade. Many—twenty yesterday—are carted out daily, who have died from unknown causes, and whom the medical officers have never seen. The dead are hauled out daily by the wagon load, and buried without coffins, their hands in many instances being first mutilated with an axe in the removal of any finger rings they may have. The sanitary condition of the prisoners is as wretched as can be, the principal causes of mortality being scurvy and chronic diarrhœa, the percentage of the former being disproportionately large among those brought from Belle Isle. Nothing seems to have been done, and but little if any effort made to arrest it by procuring proper food. The ration is one-third of a pound of bacon, and a pound and a quarter of unbolted corn meal, with fresh beef at rare intervals, and occasionally rice. When to be obtained—very seldom—a small quantity of molasses is substituted for the meat ration. A little weak vinegar, unfit for use has sometimes been issued. The arrangements for cooking and baking have been wholly inadequate, and though additions are now being completed, it will still be impossible to cook for the whole number of prisoners. Raw rations have to be issued to a very large proportion, who are entirely unprovided with proper utensils, and furnished so limited a supply of fuel they are compelled to dig with their hands in the filthy marsh before mentioned for roots, &c. No soap or clothing has ever been issued. After inquiring, I am confident that by slight exertions green corn and other anti-scorbutics could readily be obtained. I herewith hand two reports of Chief Surgeon White, to which I would respectfully call your attention. The present hospital arrangements were only intended for the accommodation of the sick of 10,000 men, and are totally insufficient, both in character and extent, for the present needs, the number of prisoners being now more than three times as great, the number of cases requiring medical treatment is in an increased ratio. It is impossible to state the number of sick, many dying within the stockade whom the medical officers never see or hear of until their remains are brought out for interment. The rate of death has been steadily increased from 374-10 per mil. during the month of March last, to 62 7-10 per mil. in July. Of the medical officers, but ten hold commissions; nearly all of the others are detailed from the militia, and have accepted the position to avoid serving in the ranks, and will relinquish their contracts as soon as the present emergency is passed and the militia is disbanded. But little injury would result from this, however, as they are generally very inefficient. Not residing at the post, only visiting it once a day at sick call, they bestow but little attention to those under their care. The small pox hospital is under the charge of Dr. E. Sheppard, P. A. C. S. More than half the cases in it have terminated fatally. The management and police of the general hospital grounds seem to be as good as the limited means will allow, but there is pressing necessity for at least three times the number of tents and amount of bedding now on hand. The supply of medicines is wholly inadequate, and frequently there is none, owing to the great delays experienced in filling the requisitions.

In conclusion, I beg leave to recommend that no more prisoners be sent to this already overcrowded prison, and that at the two additional localities selected by General Winder under instructions from General Bragg, the one near Millen, Georgia, the other some point in Alabama south of Cahawba, arrangements be at once made for the excess over 15,000 at this post, and such others as may be captured. Since my inspection was made, over 1,300 prisoners have been added to the number specified in the reports herewith. With a view of relieving to some extent this point as soon as possible, I respectfully suggest that 2,000 of those who most need the change, especially the Belle Isle prisoners, be at once sent to Macon, to occupy the quarters vacated by the Federal officers, that being the greatest number that can be properly accommodated with shelter at that point.

I am, Colonel, your obedient servant,

D. T. CHANDLER,
Assistant Adjutant and Inspector General.

Colonel R. H. CHILTON,
Assistant Adjutant and Inspector General.

APPENDIX. 5

DOCUMENT No. 2.

SURGEON ISAIAH WHITE'S REPORT TO COLONEL CHANDLER.

CHIEF SURGEON'S OFFICE, August 2, 1864.

COLONEL: I have the honor to submit the following report of the sanitary condition of the Confederate States military prison:

The number of sick on morning report is one thousand three hundred and five (1,305) in hospital, and five thousand and ten (5,010) in quarters.

The total number of deaths from the organization of the prison (February 24, 1864) up to date, is 4,585.

The following table exhibits the ratio per one thousand (1,000) of mean strength during the different months:

Month.	Mean Strength.	Deaths.	Ratio per 1,000 of Mean Strength.
March	7,500	283	37.4
April	10,000	576	57.6
May	15,000	708	47.2
June	22,291	1,201	53.87
July	29,030	1,817	62.7

Owing to insufficient hospital accommodation, many are treated in quarters who should be in hospital. The present capacity of the hospital is for 1,400 sick. The hospital is situated in an oak grove, affording good shade. Through the hospital passes a stream, furnishing an ample supply of water for cleanliness; drinking water is obtained, of good quality, from wells and springs on the banks of the stream.

The tents are insufficient in number, and not of proper size for the treatment of sick ; most of them are the small fly tent and tent flies. There should be at least two hundred hospital or five hundred wall tents to properly accommodate the sick. It has been impossible up to this time to obtain straw for bedding, this not being a grain-growing district ; small crops of wheat have been grown this year, and efforts are being made to collect a sufficient quantity as soon as the present crop is thrashed ; but there is a lack of transportation at the post, and farmers are unwilling to hire their own teams for the purpose. The attendants are paroled prisoners, who, as a rule, are faithful to the performance of their duty, being actuated by the improvement of their own condition on removal from the stockade, and a fear of a return if negligent in the performance of duty, apart from a desire to serve their own sick comrades. The number of medical officers, until the recent call of militia by the Governor of Georgia, was utterly inadequate; since that time a number of physicians have been employed by contract, and others have been detailed by the Governor to serve in the medical department. These have been recently assigned, and it is impossible to decide on their proficiency. The other medical officers, with a few exceptions, are capable and attentive. The physicians who have been recently employed will no doubt cancel their contracts as soon as the militia is disbanded, and the services of the detailed physicians will also be lost. With this view I would suggest that a sufficient number of competent medical officers be assigned.

There is a deficiency of medical supplies issued by the medical purveyor. Supplies of medicines have occasionally been entirely exhausted, and we have been left several days at a time without any whatever. This has arisen from the delay experienced in sending requisitions to medical director at Atlanta for approval.

The hospital ration is commuted as for other general hospitals, and supplies for the subsistence and comfort of sick are purchased with hospital fund. Heretofore we have been able to supply sick with vegetables: but during the entire month of July the commissary has been without funds, and difficulty has been experienced in purchasing on time.

The ration issued to the prisoners is the same as that issued to the Confederate soldiers in the field, viz.: one-third of a pound of pork, and a pound and a quarter of meal, with an occasional issue of beans, rice, and molasses.

The meal is issued unbolted, and when baked is coarse and unwholesome.

Amongst the old prisoners, scurvy prevails to a great extent, which is usually accompanied by diseases of the digestive organs. This, in connection with the mental depression produced by long imprisonment, is the chief cause of mortality. There is nothing in the topography of the country that can be said to influence the health of the prison.

The land is high and well drained, the soil light and sandy, with

APPENDIX. 7

no marshes or other source of malaria in the vicinity, except the small stream within the stockade. The densely crowded condition of the prisoners, with the innumerable little shelters irregularly arranged, precludes the enforcement of proper police, and prevents free circulation of air.

The lack of barrack accommodation exposes the men to the heat of the sun during the day and to the dew at night, and is a prolific source of disease.

The margins of the stream passing through the stockade are low and boggy, and having been recently drained, have exposed a large surface covered with vegetable mould to the rays of the sun, a condition favorable to the development of malarious diseases. It is the design of the commandant of the prison to cover the surface with dry sand, but the work has been unavoidably retarded.

The absence of proper sinks (and the filthy habits of the men) have caused a deposit of fecal matter over almost the entire surface of this bottom land.

The point of exit of the stream through the walls of the stockade is not sufficiently bold to permit a free passage of ordure.

When the stream is swollen by rains the lower portion of this bottom land is overflowed by a solution of excrement, which, subsiding, and the surface exposed to the sun, produces a horrible stench.

Captain Wirz, the commandant of the prison, has doubtless explained to you the difficulties which have prevented these, with other projected improvements, in the way of bathing and other arrangements for cleanliness.

Respectfully submitted :

ISAIAH H. WHITE,
Chief Surgeon Post.

Colonel CHANDLER.

DOCUMENT No. 3.

FROM THE TESTIMONY OF FATHER HAMILTON.

The priests who went there after me, while administering the sacrament to the dying, had to use an umbrella, the heat was so intense. Some of them broke down in consequence of their services there. In the month of August, I think, we had three priests there constantly. We had a priest from Mobile, who spoke three or four languages, inasmuch as you could find every nationality inside the stockade, and two from Savannah, and we had one from Augusta at another time. One of the priests from Savannah came to Macon, where I reside, completely prostrated, and was sick at my house for several days.

As I said before, when I went there, I was kept so busily engaged in giving the sacrament to the dying men, that I could not observe much; but of course I could not keep my eyes closed as to what I saw there. I saw a great many men perfectly naked, walking about through the stockade perfectly nude; they seemed to have lost all regard for delicacy, shame, morality or anything else. I would frequently have to creep on my hands and knees into the holes that the men had burrowed in the ground and stretch myself out alongside of them to hear their confessions. I found them almost living in vermin in those holes; they could not be in any other condition than a filthy one, because they got no soap and no change of clothing, and were there all huddled up together.

I never at any time counted the number of dead bodies being taken out of the stockade in the morning. I have never seen any dead carried out of the stockade. I have seen dead bodies in the hospital in the morning. In the case of the man in the hospital of whom I was speaking a while ago, after I had heard his confession, and before I gave him the last rites of the church sacrament in "extreme unction," as we call it, I saw them placing the night guards in the hospital, and knew that I would not be able to get out after that. I told him that I would return in the morning and give him the other rites of the church, if he still lived. I was in there early the next morning, and in going down one of the avenues I counted from forty to sixty dead bodies of those who had died during the night in the hospital. I had never seen any dead bodies in the stockade. I have seen a person in the hospital in a nude condition, perfectly naked. They were not only covered with the ordinary vermin, but with maggots. They had involuntary evacuations, and there were no persons to look after them. The nurses did not seem to pay any attention whatever, and in consequence of being allowed to lie in their own filth for some hours, vermin of every description had got on them, which they were unable to keep off them. This was in the latter part of May. I never noticed in the stockade the men digging in the ground, and standing in the sand to protect themselves from the sun. I did not see any instance of that kind. I have seen them making little places from a foot to a foot and a half deep, and stretching their blankets right over them. I have crawled into such places frequently to hear the confessions of the dying. They would hold from one to two; sometimes a prisoner would share his blanket with another, and allow him to get under shelter.

DOCUMENT No. 4.

REPORT OF SURGEON JONES, C. S. A.

The Confederate military prison at Andersonville, Georgia, consists of a strong stockade, twenty feet in height, enclosing twenty-seven acres. The stockade is formed of strong pine logs, firmly planted in the ground. The main stockade is surrounded by two other similar rows of pine logs, the middle stockade being sixteen feet high, and the outer twelve feet. These are intended for offence and defence. If the inner stockade should at any time be forced by the prisoners, the second forms another line of defence; while in case of an attempt to deliver the prisoners by a force operating upon the exterior, the outer line forms an admirable protection to the Confederate troops, and a most formidable obstacle to cavalry or infantry. The four angles of the outer line are strengthened by earthworks upon commanding eminences, from which the cannon, in case of an outbreak among the prisoners, may sweep the entire enclosure; and it was designed to connect these works by a line of rifle pits, running zigzag around the outer stockade; those rifle pits have never been completed. The ground enclosed by the innermost stockade lies in the form of a parallelogram, the larger diameter running almost due north and south.

The stockade was built originally to accommodate only 10,000 prisoners, and included at first seventeen acres. Near the close of the month of June, the area was enlarged by the addition of ten acres. The ground added was situated on the northern slope of the largest hill.

Within the circumscribed area of the stockade, the Federal prisoners were compelled to perform all the offices of life—cooking, washing, urinating, defecation, exercise and sleeping. During the month of March the prison was less crowded than at any subsequent time, and then the average space of ground to each prisoner was only 98.7 feet, or less than seven square yards. The Federal prisoners were gathered from all parts of the Confederate States east of the Mississippi, and crowded into the confined space, until in the month of June the average number of square feet of ground to each prisoner was only 33.2, or less than four square yards. These figures represent the condition of the stockade in a better light even than it really was; for a considerable breadth of land along the stream, flowing from west to east, between the hills, was low and boggy, and was covered with the excrement of the men, and thus rendered wholly uninhabitable, and, in fact, useless for every purpose except that of defecation. The pines and other small trees and shrubs, which originally were scattered sparsely over these hills, were in a short time cut down and consumed by the prisoners for firewood, and no shade tree was left in the entire enclosure of the stockade. With their characteristic industry and ingenuity, the Federals constructed for themselves small huts and caves, and attempted to shield themselves from the rain and sun and night damps and dew. But few tents were distributed to the prisoners, and those were in most cases torn and rotten.

THE HOSPITAL.

The entire grounds are surrounded by a frail board fence, and are strictly guarded by Confederate soldiers, and no prisoner except the paroled attendants is allowed to leave the grounds except by a special permit from the commandant of the interior of the prison.

The patients and attendants, near two thousand in number, are crowded into this confined space and are but poorly supplied with old and ragged tents. Large numbers of them were without any bunks in the tents, and lay upon the ground, oftentimes without even a blanket. No beds or straw appeared to have been furnished. The tents extend to within a few yards of the small stream, the eastern portion of which, as we have before said, is used as a privy and is loaded with excrements; and I observed a large pile of corn bread, bones and filth of all kinds, thirty feet in diameter and several feet in height, swarming with myriads of flies, in a vacant space near the pots used for cooking. Millions of flies swarmed over everything and covered the faces of the sleeping patients, and crawled down their open mouths, and deposited their maggots in the gangrenous wounds of the living and in the mouths of the dead. Mosquitos in great numbers also infested the tents, and many of the patients were so stung by these pestiferous insects, that they resembled those suffering with a slight attack of the measles.

The manner of disposing of the dead was also calculated to depress the already desponding spirits of these men, many of whom have been confined for months, and even for near two years in Richmond and other places, and whose strength has been wasted by bad air, bad food, and neglect of personal cleanliness. The dead house is merely a frame covered with old tent cloth and a few bushes, situated in the southwestern corner of the hospital grounds. When a patient dies he is simply laid in the narrow street in front of his tent, until he is removed by Federal negroes detailed to carry off the dead; if a patient dies during the night, he lies there until the morning, and during the day, even, the dead were frequently allowed to remain for hours in these walks. In the dead house the corpses lie upon the bare ground, and were in most cases covered with filth and vermin.

There appeared to be almost absolute indifference and neglect on the part of the patients of personal cleanliness; their persons and clothing in most instances, and especially of those suffering with gangrene and scorbutic ulcers, were filthy in the extreme and covered with vermin. It was too often the case that patients were received from the stockade in a most deplorable condition. I have seen men brought in from the stockade in a dying condition, begrimed from head to foot with their own excrements, and so black from smoke and filth that they resembled negroes rather than white men. That this description of the stockade and hospital has not been overdrawn, will appear from the reports of the surgeons in charge, appended to this report.

DOCUMENT No. 5.

DR. PELOT'S REPORT.

FIRST DIVISION, C. S. M. P.
HOSPITAL, September 5, 1864.

SIR: As officer of the day for the past twenty-four hours, I have inspected the hospital and found it in as good condition as the nature of the circumstances will allow. A majority of the bunks are still unsupplied with bedding, while in a portion of the division the tents are entirely destitute of either bunks, bedding or straw, the patients being compelled to lie upon the bare ground. I would earnestly call attention to the article of diet. The corn bread received from the bakery being made up without sifting, is wholly unfit for the use of the sick; and often (in the last twenty-four hours) upon examination the inner portion is found to be perfectly raw. The meat (beef) received by the patients does not amount to over two ounces a day, and for the past three or four days no flour has been issued. The corn bread cannot be eaten by many, for to do so would be to increase the diseases of the bowels, from which a large majority are suffering, and it is therefore thrown away. All their rations received by way of sustenance is two ounces of boiled beef and half pint of rice soup per day. Under these circumstances, all the skill that can be brought to bear upon their cases by the medical officer will avail nothing. Another point to which I feel it my duty to call your attention, is the deficiency of medicines. We have but little more than indigenous barks and roots with which to treat the numerous forms of disease to which our attention is daily called. For the treatment of wounds, ulcers, &c., we have literally nothing except water.

Our wards—some of them—were filled with gangrene, and we are compelled to fold our arms and look quietly upon its ravages, not even having stimulants to support the system under its depressing influences, this article being so limited in supply that it can only be issued for cases under the knife. I would respectfully call your earnest attention to the above facts, in the hope that something may be done to alleviate the sufferings of the sick.

I am, sir, very respectfully, your obedient servant,

J. CREWS PELOT,
Assistant Surgeon C. S. and Officer of the Day.

APPENDIX. 13

DOCUMENT No. 6.

CONSOLIDATED RETURN FOR CONFEDERATE STATES MILITARY PRISON, CAMP SUMTER, ANDERSONVILLE, GEORGIA, FOR THE MONTH OF AUGUST, 1864.

Prisoners on hand 1st of August, 1864:
In camp.. 29,985
In hospital.................................... 1,693
                 31,678
Received from various places during August....... 3,078
Recaptured 4   3,082
 Carried out 34,760
Died during the month of August................ 2,993
Sent to other parts............................ 23
Exchanged 21
Escaped........................... 30   3,067   3,061
Total on hand................................ 31,693

Of which there are on the 31st of August—
In camp....................................... 29,473
In hospital 2,220
                 31,693

DOCUMENT No. 7.

DR. HOPKINS' REPORT.

ANDERSONVILLE, GEORGIA, August 1, 1864.

GENERAL: In obedience to your order of July 28, requiring us to make a careful examination of the Federal prison and hospital at this place, and to ascertain and report to you the cause of disease and mortality among the prisoners, and the means necessary to prevent the same, this has been complied with, and we respectfully submit the following:

CAUSE OF DISEASE AND MORTALITY.

1. The large number of prisoners crowded together.
2. The entire absence of all vegetables as diet, so necessary as a preventive of scurvy.

3. The want of barracks to shelter the prisoners from sun and rain.
4. The inadequate supply of wood and good water.
5. Badly cooked food.
6. The filthy condition of prisoners and prison generally.
7. The morbific emanations from the branch or ravine passing through the prison, the condition of which cannot be better explained than by naming it a morass of human excrement and mud.

PREVENTIVE MEASURES.

1. The removal immediately from the prison of not less than 15,000 prisoners.
2. Detail on parole a sufficient number of prisoners to cultivate the necessary supply of vegetables, and until this can be carried into practical operation, the appointment of agents along the different lines of railroads to purchase and forward a supply.
3. The immediate erection of barracks to shelter the prisoners.
4. To furnish the necessary quantity of wood, and have wells dug to supply the deficiency of water.
5. Divide the prisoners into squads, place each squad under the charge of a sergeant, furnish the necessary quantity of soap, and hold these sergeants responsible for the personal cleanliness of his squad; furnish the prisoners with clothing at the expense of the Confederate government, and if that government be unable to do so, candidly admit our inability and call upon the Federal government to furnish them.
6. By a daily inspection of bakehouse and baking.
7. Cover over with sand from the hillsides the entire "morass" not less than six inches deep, board the stream or watercourse, and confine the men to the use of the sinks, and make the penalty for disobedience of such orders *severe*.

FOR THE HOSPITAL.

We recommend—
1st. The tents be floored with planks; if plank cannot be had, with puncheons; and if this be impossible, then with fine straw, to be frequently changed.
2d. We find an inadequate supply of stool boxes, and recommend that the number be increased, and that the nurses be required to re-

move them as soon as used, and before returning them see that they are well washed and limed.

3d. The diet for the sick is not such as they should have, and we recommend that they be supplied with the necessary quantity of beef soup with vegetables.

4th. We also recommend that the surgeons be required to visit the hospitals not less than twice a day.

We cannot too strongly recommend the necessity for the appointment of an efficient medical officer to the exclusive duty of inspecting daily the prison hospital and bakery, requiring of him daily reports of their condition to headquarters.

We have the honor to remain, general, very respectfully,

T. S. HOPKINS,
Acting Assistant Surgeon.

DOCUMENT No. 8.

TESTIMONY OF BOSTON CORBETT.

It was a living mass of putrefaction and filth; there were maggots there a foot deep. Any time we turned over the soil we could see the maggots in a living mass; I have seen the soldiers wading through it, digging for roots to use for fuel. I have seen around the swamp, the sick in great numbers, lying pretty much as soldiers lie when they are down to rest in line after a march. In the morning I could see those who had died during the night, and in the daytime I could see them exposed to the heat of the sun, with their feet swelled to an enormous size; in many cases large gangrene sores filled with maggots and flies which they were unable to keep off. I have seen men lying there in a state of utter destitution, not able to help themselves, lying in their own filth. They generally chose that place (near the swamp), those who were most offensive, because others would drive them away, not wanting to be near those who had such bad sores. They chose it because of its being so near to the sinks. In one case a man died there, I am satisfied, from the effects of lice. When the clothes were taken off his body, the lice seemed as thick as the garment—a living mass.

DOCUMENT No. 9.

DR. M. M. MARSH'S TESTIMONY AS TO STORES SENT TO PRISONERS AT ANDERSONVILLE, GEORGIA.

5,052 wool shirts.
6,993 wool drawers.
3,950 handkerchiefs.
601 cotton shirts.
1,128 cotton drawers.
2,100 blouses.
4,235 wool pants.
1,520 wool hats.
2,565 overcoats.
5,385 blankets.
272 quilts.
2,120 pairs shoes.
110 cotton coats.
140 vests.
46 cotton pants.
534 wrappers.
69 jackets.
12 overalls.
817 pairs slippers.
3,147 towels.

5,431 wool socks.
50 pillow cases.
258 bed sacks.
122 combs.
100 tin cups.
2 boxes tinware.
4,092 pounds condensed milk.
4,032 pounds condensed coffee.
1,000 pounds farina.
1,000 pounds corn starch.
4,212 pounds tomatos.
24 pounds chocolate.
3 boxes lemon juice.
1 barrel dried apples.
111 barrels crackers,
60 boxes cocoa.
7,200 pounds beef stock,
Paper, envelopes, &c.
Pepper, mustard.
One box tea, 70 pounds.

DOCUMENT No. 10.

(From Private Diary of J. W. Merrill.)

MARKET PRICES IN C. S. M. PRISON HOSPITAL, ANDERSONVILLE, GEORGIA, 1864.

To Sept. #4 Confed. — $1 Federal.	Federal Money.	Confederate Money.	After Sept. #5 Confed. — $1 Federal.	Federal Money.	Confederate Money.
	$ c. $ c.	$ c. $ c.		$ c. $ c.	$ c. $ c.
Apples, each....	0 10 to 0 25	0 40 to 1 00	Onions, each.....	0 25 to 0 50	1 00 to 2 00
Butter, lb.......	1 50	6 00	Peanuts, pint....	0 30	1 50
Biscuit. each....	0 08 to 0 15	0 32 to 0 60	Potatoes, Irish, qt	1 00	5 00
Blackberries, p't	0 50 to 0 75	2 00 to 3 00	Potat's, Sw't, ea.	0 10 to 0 15	0 50 to 0 75
Whortleberries.	0 75	3 00	Peaches, each...	0 10 to 0 75	0 40 to 3 00
Chestnuts, each.	0 01	0 05	Sugar Cane Stalk	0 25	1 25
Chincopias, pint	1 00	5 00	Red Peppers, ea.	0 10	0 50
Eggs, each......	0 15 to 0 25	0 60 to 1 00			
Grapes, each....	0 01	0 05			
Gingerbread, ea.	0 35 to 0 50	1 40 to 2 00			
Honey, table-spoonful......	0 18	0 50			
Molasses, quart.	3 00	15 00			
Watermelons, ea	1 50 to 3 00	6 00 to 12 00			
Muskmelons, ea	0 50 to 1 50	2 00 to 6 00			

DOCUMENT No. 11.

(From Private Diary of J. W. Merrill.)

MINUTES OF A MEETING OF THE SERGEANTS IN CHARGE OF DETACHMENTS OF PRISONERS CONFINED AT ANDERSONVILLE, GA.

At a meeting of the sergeants in charge of the various detachments of prisoners confined at present in the Andersonville Military Prison, Georgia, held for the purpose of taking some action to properly represent the present condition of the prisoners to our Government at Washington, and thereby secure, if possible, a speedy redress of the wrongs complained of, the following committee was appointed, who, after due consultation, reported the following preamble and resolutions, which were unanimously adopted:

William N. Johnson, *Chairman.*

H. C. Higginson, J. S. Banks, E. W. Webb, *Committee.*

Apparently, one of the sad effects of the progress of this terrible war has been to deaden our sympathies and make us more selfish than we were when the tocsin of the battle strife first sounded in the land.

Perhaps this state of public feeling was to have been anticipated. The frequency with which you hear of captures in battle, and the long accounts you have seen of their treatment, has robbed the spectacle of its novelty, and, by a law of our nature, has taken off the edge of our sensibilities, and made them less an object of interest. No one can know the horrors of imprisonment in crowded and filthy quarters but he who has endured it, and it requires a brave heart not to succumb. But hunger, filth, nakedness and disease are as nothing compared with that heart sickness which weighs prisoners down, most of them young men whose terms of enlistment have expired, and many of them with nothing to attach them to the cause in which they suffer but principle and love of country and of friends. Does the misfortune of being taken prisoners make us less the object of interest and value to our Government ? If such, you plead it no longer. These are no common men, and it is no common merit that they call upon you to aid in their release from captivity.

1st. That a large portion of the prisoners have been held as such for periods ranging from nine to fifteen months, subject to all the hardships and privations necessarily incident to a state of captivity in an enemy's country.

2d. That there are now confined in the prison from 25,000 to 30,000 men, with daily accession of hundreds, and that the mortality among them, generated by various causes, such as change of climate, diet and want of proper exercise, is becoming truly frightful to contemplate, and is rapidly increasing in virulence, decimating their ranks weekly by hundreds.

3d. In view of the foregoing facts, we, your petitioners, most earnestly yet respectfully pray, that some action be immediately taken to effect our speedy release, either on parole or exchange, the dictates both of justice and humanity alike demanding such action on the part of our Government.

4th. We shall all look forward with a hopeful confidence that something will be speedily done in this matter, believing that a proper statement of the facts is all that is necessary to secure a redress of the grievances complained of.

5th. The above has been read to each detachment by its respective sergeant, and has been approved by the men, who have unanimously authorized each sergeant to sign, as will and deed of the whole.

ANDERSONVILLE, GEO.

DOCUMENT No. 12.

(*From Private Diary of J. W. Merrill.*)

MORTALITY AT C. S. M. PRISON HOSPITAL DURING THE MONTHS OF AUGUST, SEPTEMBER AND OCTOBER, 1864.

AUG.		SEPT.		OCT.	
1	74	1	105	1	82
2	73	2	104	2	48
3	75	3	113	3	40
4	75	4	94	4	66
5	90	5	98	5	46
6	103	6	105	6	49
7	71	7	63	7	53
8	95	8	111	8	52
9	95	9	76	9	34
10	85	10	99	10	64
11	103	11	99	11	103
12	81	12	111	12	76
13	109	13	78	13	60
14	114	14	102	14	54
15	120	15	83	15	47
16	107	16	100	16	51
17	114	17	106	17	48
18	88	18	129	18	53
19	101	19	90	19	55
20	107	20	99	20	41
21	86	21	82	21	41
22	122	22	61	22	50
23	127	23	82	23	51
24	102	24	77	24	67
25	98	25	72	25	22
26	103	26	51	26	71
27	93	27	83	27	40
28	90	28	75	28	37
29	105	29	69	29	28
30	95	30	60	30	39
31	92			31	27
Total	2,993	Total	2,677	Total	1,595

DOCUMENT No. 13.

(*From Private Diary of J. W. Merrill.*)

REPORT OF ELECTION HELD IN ANDERSONVILLE PRISON HOSPITAL, TUESDAY, NOV. 8, 1864.

(*Written by S. M. Riker.*)

On the evening of November 7th, 1864, the prisoners of the C. S. Military Prison Hospital, Andersonville, Ga., held a meeting for the discussion of the opposing candidates for the office of President of the United States for the ensuing four years, having first obtained permission from Capt. Wirz, Military Commander of Prison, to carry on an informal election in the prison.

On motion, Mr. Smith was voted chairman of the meeting. Other offices, owing to existing circumstances, were ignored.

The meeting was opened by singing the national anthem, "Columbia, the Gem of the Ocean," conducted by Mr. J. W. Merrill; after which Dr. J. W. Fay mounted the rostrum, and edified the meeting by an able address. He held that the present administration, having sworn, upon entering upon its duties, to carry out the laws and Constitution of the United States, had been faithful to its oath, acting wisely and just in every emergency. He considered the fealty of General McClellan to the Union very doubtful, owing to his conduct after the battle of Antietam and his intercourse and association with such traitors and copperheads as Judge Woodward of Pennsylvania, Vallandigham, Hall, Wood Brothers, &c., of the Chicago Convention.

The Doctor was replied to by S. M. Riker in a few remarks, sustaining the character of General McClellan for truthfulness and faithfulness to his country and the cause for which he had fought. Mr. Riker then made a long, animated speech. He was loudly applauded, the conciseness of his arguments and palpableness of his conclusions having a telling effect on his hearers.

Mr. Delaney replied to Mr. Riker in a very terse manner, denouncing Old Abe with his greenbacks, and establishing, beyond a doubt, that his education and politics were of the real "hard shell" democratic order.

Mr. Lathrop being then called for, took the stand and addressed the meeting for about fifteen minutes in an easy, eloquent manner. He upheld the present administration, and was for carrying on this war to the bitter end. He would advocate no treaty of peace that would in any manner compromise the people of the North, but would, if necessary, enlist for four, eight or sixteen years—longer, if necessary—to conquer the rebels.

Mr. Waterbury took the stand. He claimed that the nigger was prized more highly by the present administration than the white soldier—that a dozen niggers were the cause of the Government's ceasing to exchange prisoners, and niggers were the cause of our now being prisoners in the Confederacy. He considered he had been deceived by the Government, and that the soldiers of our armies were, as a whole, swindled.

Loud calls were made for Mr. Bennett, who came forward, and, smiling all over, proclaimed for honest Old Abe. He spoke for some twenty minutes in a very racy and humorous strain, not without effect, as the frequent bursts of applause gave abundant evidence. He admitted General McClellan to be a man of fine intellect and abilities, but much preferred to ride the old horse, who was still able to carry him, and had never yet stumbled. He considered the anaconda business of little Mac played out. This allowing the body to lie in the way of the advancing rebels, while the tail swings around and crushes them, was not the one that would elect him to the highest office in the favor of the people of the United States. The anaconda game was a thorough fizzle, in his estimation.

Mr. B. resumed his seat amidst thunders of applause, and was followed by Mr. Burns, who edified the audience with a reviewal of the arguments of his predecessor, deducing therefrom an argument in favor of Mr. Lincoln.

The Chairman of the meeting closed with some pointed arguments favoring General McClellan for the next President of the United States.

Throughout the evening, the choir, under the direction of Mr. J. W. Merrill, furnished the meeting with fine singing. "The Star Spangled Banner," "Hail Columbia," "Home, Sweet Home," "Rally Round the Flag, Boys," "Hoist up the Flag," and others, were rendered with great taste and effect, and received by the audience with much applause.

Mr. William West sang the "American Star," when the meeting adjourned.

All were highly gratified with the evening's performance. The greatest harmony and good feeling existed.

The Committee congratulate themselves on the attainment of the object of the meeting, namely, amusement. Mingled with amusement was an air of earnestness which did credit to the assembly, as citizens of America, and evidenced the degree of interest that they felt in a matter of such vital importance to the country for which they are sworn to fight. Each and every one seemed to feel, that though their votes could have no visible effect on the struggle of the next day, in which the whole loyal population of the North were to engage, yet the principles involved and expressed would be the same.

On the 8th inst. an election for President was held. Mr. M. E. Hogan, of Third Indiana Cavalry, Magistrate.

Judges of Election—First Division : John Dunmore, One Hundredth Ohio ; A. A. Walker, Sixteenth Connecticut. Second Division : E. H. Lathrop, Eighty-first Illinois ; William Smith, Sixth Michigan. Third Division : T. M. Seaton, Eighteenth New York ; John Cornwall. Fourth Division : J. W. Merrill, Twenty-fourth New York Battery ; Charles Dunmore, Sixteenth Illinois Cavalry.

The day was rainy, and just the kind for an election, which went off in a very satisfactory manner.

At six P. M. the result was announced, the camp having given a vote of 150 majority for Abraham Lincoln for President of the United States. 1,740 votes in all were polled ; Lincoln receiving 945, and McClellan 795.

www.ingramcontent.com/pod-product-compliance
Lightning Source LLC
Chambersburg PA
CBHW030748230426
43667CB00007B/889